AF539390

SOIL ENGINEERING
TESTING DESIGN AND REMEDIATION

SOIL ENGINEERING
TESTING DESIGN AND REMEDIATION

Ravindra Singh

NEW DELHI - 110 002 (INDIA)

Soil Engineering: Testing Design and Remediation

ISBN 978-93-51117-09-4

Published in 2015 in India by

RANDOM PUBLICATIONS

4376-A/4B, Gali Murari Lal, Ansari Road
New Delhi-110 002
Phone: +9111-43580356, 23289044
E-mail: randomexports@gmail.com; sales@randompublications.com; info@randompublications.com

Reprinted 2025

Type Setting by: Friends Media, Delhi-110089
Digitally Printed at : Replika Press Pvt. Ltd.

Preface

Humans have historically used soil as a material for flood control, irrigation purposes, burial sites, building foundations, and as construction material for buildings. First activities were linked to irrigation and flood control, as demonstrated by traces of dykes, dams, and canals dating back to at least 2000 BCE that were found in ancient Egypt, ancient Mesopotamia and the Fertile Crescent, as well as around the early settlements of Mohenjo Daro and Harappa in the Indus valley. As the cities expanded, structures were erected supported by formalized foundations; Ancient Greeks notably constructed pad footings and strip-and-raft foundations. Until the 18th century, however, no theoretical basis for soil design had been developed and the discipline was more of an art than a science, relying on past experience.

Geotechnical engineering is the branch of civil engineering concerned with the engineering behaviour of earth materials. Geotechnical engineering is important in civil engineering, but also has applications in military, mining, petroleum and other engineering disciplines that are concerned with construction occurring on the surface or within the ground. Geotechnical engineering uses principles of soil mechanics and rock mechanics to investigate subsurface conditions and materials; determine the relevant physical/mechanical and chemical properties of these materials; evaluate stability of natural slopes and man-made soil deposits; assess risks posed by site conditions; design earthworks and structure foundations; and monitor site conditions, earthwork and foundation construction. In geotechnical engineering, soils are considered a three-phase material composed of: rock or mineral particles, water and air. The voids of a soil, the spaces in between mineral particles, contain the water and air. The engineering properties of soils are affected by four main factors: the predominant size of the mineral particles, the type of mineral particles, the grain size distribution, and the relative quantities of mineral, water and air present in the soil matrix. Fine particles (fines) are defined as particles less than 0.075 mm in diameter. A typical geotechnical engineering project begins with a review of project needs to define the required material properties. Then follows a site investigation of soil, rock, fault distribution and bedrock properties on and below an area of interest to determine their engineering properties including how they will interact with, on or in a proposed construction. Site investigations are needed to gain an understanding of the area in or on which the engineering will take place. Investigations can include the assessment of the risk to humans, property and the environment from natural hazards such as earthquakes, landslides, sinkholes, soil liquefaction, debris flows and rockfalls. Ground Improvement refers to a technique that improves the engineering properties of the soil mass treated. Usually, the properties that are modified are shear strength, stiffness and permeability. Ground improvement has developed into a sophisticated tool to support foundations for a wide variety of structures. Properly applied, i.e. after giving due consideration to the nature of the ground being improved and the type and sensitivity of the structures being built, ground improvement often reduces direct costs and saves time. A geotechnical engineer then determines and designs the type of foundations, earthworks, and/or pavement subgrades required for the intended man-made structures to be built. Foundations are designed and constructed for structures of various sizes such as high-rise buildings, bridges, medium to large commercial buildings, and smaller structures where the soil conditions do not allow code-based design. Foundations built for above-ground structures include shallow and deep foundations. Retaining structures include earth-filled dams and retaining walls. Earthworks include embankments, tunnels, dikes and levees, channels, reservoirs, deposition of hazardous waste and sanitary landfills.

This book provides the practical meaning of the different aspects of soil mechanics, the use of unconfined compression test data, the meaning of consolidated tests, the practical value of lateral pressure, and more. Though the book is basically meant for the student, the nature, content and manner of presentation of the material makes it a useful reference book for any practitioner of the art of Soil Engineering.

I thank all members of my team who have helped in the preparation of the book. My special thanks go to "Random Publications" who have published the book.

— *Ravindra Singh*

Contents

Chapter 1

Functional Groups of Soil Organisms

Different soil organisms are able to carry out the same functions in soil. This applies to common processes such as the release of ammonium from organic matter and the conversion of ammonium to nitrate. This is because many soil organisms perform similar functions, and are collectively called 'functional groups'. These groups can be quantified. One method for quantifying the number of functional groups of soil organisms is to determine the 'microbial catabolic diversity' of soils. This approach quantifies the microbial population in the soil based on its ability to degrade a wide selection of organic carbon molecules, which are added separately to the soil. The amount of soil respiration is measured after the addition of each carbon molecule and this is used to indicate the level of degradation. The pattern of degradation of these substrates corresponds with the expected activities of organisms in soil (Degens and Harris 1996). An alternative method measures enzyme activity in response to the addition of a range of carbon substrates to a solution containing organisms washed out of the soil.

Quantifying the functional groups of organisms in soil involved in specific processes is an important research area as it provides knowledge about the genetic diversity of soil communities. Understanding the diversity of soil organisms may be for important for questions about the sustainability of land use.

Relationships Between Abundance of Different Types of Organisms

Close links exist between the abundance of some groups of organisms in the soil and in some instances, it is possible to use an assessment of the abundance of one group of organisms to estimate the abundance of another. An example of this is the strong correspondence between the number of fungus-eating amoeba and the quantity of fungal hyphae in soil (Gupta and Germida 1988). This reflects the strong link between the quantity of amoebae and their food source (hyphae).

The relative abundance of fungi and bacteria can also be estimated as an indication of the status of the biological component of soil. Generally, a higher ratio of fungi to bacteria is indicative of more stable soil communities, whereas bacterial dominance is more common in intensively managed agricultural soils.

Why is it Useful to Use more than one Technique to Quantify Soil Organisms

Quite often, different techniques produce quite different estimates of the number of soil organisms for the same soil. For example, direct observation and plate counting can produce widely different estimates of the number of bacteria (Russell 1973). The number of bacteria counted using the plate-count method is usually only about 1% of the number that is counted by directly counting bacteria extracted from soil and collected on a filter. Two reasons for this are (i) that many of the bacteria will not grow on the artificial media used in plate counting, and (ii) that not all bacteria isolated on a filter would have been alive in the soil at the time of sampling.

The difference between the direct and indirect methods for assessing soil organisms is particularly important when making comparisons between different soil management treatments. Adding manure increases the abundance of bacteria in an alkaline soil, but decreases abundance in a more acid soil (Russell 1973). This effect was much greater when plate counts of bacteria were used than when direct observation counts were made.

This example demonstrates why it is necessary to have a good understanding of:

- the reliability of different methods for counting soil organisms,
- the ecology of the organisms being assessed, and
- how this information is to be used.

Molecular tools are now also used for identifying or estimating the abundance of a particular organism in soil (e.g. Griffiths *et al.* 2003). Detailed knowledge of each technique and the assumptions contained within them is essential for interpreting the results of assessment of the abundance of soil organisms. Furthermore, it also needs to be emphasised that every method of estimating the number of organisms in a soil sample is, at best, only an approximation of the real number that is present.

Activities of Soil Biota

What is Activity?

Most soil organisms are inactive for long periods throughout their life. For example, only about 10-15% of the bacteria in a farm or garden soil are in their most active state at any one time. Their activity is related to soil conditions: if the conditions are not suitable, organisms are inactive. Since soil conditions are rarely suitable, organisms spend long periods being inactive. Often the conditions that suit one group of organisms are different to those that suit another group, so as the soil environment changes there is a succession of activity.

Activity is a physiological state where cellular functions allow the organism to grow and reproduce. In less active states, an organism can maintain a very low level of physiological function. When conditions become suitable for the organism to grow, rapid increases in activity occur. It is under these conditions that soil organisms colonise plant roots or organic matter, release enzymes into soil and move into their reproductive states.

It is not easy to tell from simple observation whether fungi or bacteria are in an active state. When hyphae are removed from the soil, they often look the same irrespective of whether they are alive or dead. Therefore, staining methods have been developed that indicate whether hyphae are alive or not. The methods detect the presence of molecules, such as enzymes, that are only present in living organisms. Fluorescein diacetate is a chemical that stains the components of the cell where specific enzymes are active. Using this approach, the activity of fungi in straw was monitored (Wessen and Berg 1986). The proportion of active hyphae, ie the hyphae with enzymes was high 170 days after the fungi colonised the straw, but after 200 days activity levels were declining and by 300 days, there was very little activity. This decline occurred as the fungi degraded most of the molecules in the straw that were accessible.

A novel technique has been developed to determine the proportion of living bacteria in a soil. The method is based on the observation that the chemical nalidixic acid prevents the normal division of bacterial cells. Usually, bacteria divide during their reproductive phase. Nalidixic acid prevents cell division but does not stop the cells from growing.

Therefore, elongated cells are formed which grow more than twice the length of normal cells. The elongated cells are alive and the non-elongated cells are either alive but unable to grow in the soil extract or are dead. This method was applied to a soil sample to estimate the proportion of living bacteria (Bottomley and Maggard 1990). About 70% of bacteria elongated in the presence of nalidixic acid in this study. The nalidixic acid method was also used to estimate the number of a specific group of bacteria (rhizobia). In this case, the proportion of rhizobial cells stained using the fluorescent antibody technique that elongated in the presence of nalidixic acid was estimated.

Another method for assessing the activity of organisms in soil is to measure the amount of CO_2 released over several days. The process of CO_2 release is called respiration. A highly active soil community usually respires more, releasing more CO_2 than an inactive community. However, the abundance of microbial biomass in the soil and the amount of CO_2 released is not always closely related. This is because the amount of CO_2 released by soil organisms depends on factors that influence their activity. One factor that affects respiration is temperature. In soil, higher respiration occurs at higher temperatures.

The activity of enzymes produced by soil organisms can also be measured. As noted earlier, some enzymes are released into soil and remain active even after the organisms that produced them have died. Other enzymes are only active inside a living organism.

When organic material is added to a soil, the activity (measured as respiration, enzyme activity etc.) of many organisms increases rapidly. Once all the readily available material is degraded, activity returns to the original level.

Heterotrophic organisms increase rapidly in number in response to the addition of suitable carbon substrates. Autotrophic organisms do not respond directly to the addition of organic matter because they do not use it as a source of energy and carbon. However, some autotrophic organisms, such as nitrifying bacteria, increase in number in response to the release of an inorganic molecule such as ammonium from organic matter. Therefore, addition of organic matter to soil can indirectly affect the activity of some organisms even if they do not use it as a source of energy or carbon

The abundance and activity of organisms in a soil can be related to the total mass of organic matter to give a baseline for measuring any changes with land management practices. For example, the ratio of the quantity of microbial biomass to the total organic matter in soil has been used to indicate the impact of management practices on soil carbon (Sparling 1992; Wardle and Ghani 1995).

Microbial biomass C (g per g soil)

Total organic matter C (g per g soil)

Another ratio takes into account the level of activity of the soil microbial biomass in relation to the size of the microbial biomass. This second ratio is called q CO_2 or the respiratory quotient.

Respiration (g CO_2 produced per g soil per hour)

Biomass of organisms (g biomass per g soil)

Although measures of this kind provide additional information about the activities of soil organisms, caution is required in interpreting them. For example, the q CO_2 ratio increases or decreases as the soil environment changes. Further still, activity can be high even when there are only a few organisms, if there is severe stress on the organisms as might occur in a highly degraded soil environment, although this is not always so. However, it is difficult to find a single factor that reflects the functioning of the community of organisms in soil.

What do Organisms do in Soil?

An overview of the processes that soil organisms are involved in is presented here as an introduction to the diversity of activities of soil organisms. In Parts 2 and 3 of this book, each activity is covered in more detail. Biological processes occur concurrently or sequentially in soil; they are not isolated occurrences.

Soil organisms are involved in important processes such as:

- soil formation and aggregation
- organic matter breakdown
- degradation of toxic substances
- transformation of inorganic molecules
- beneficial associations with plants
- development and prevention of plant disease

Each of these processes is discussed briefly below.

Soil Formation and Aggregation

Some microorganisms contribute to the formation of soil, which develops from the erosion of parent rock. The influence of lichens on rock surfaces provides a good example of a microbial contribution to soil formation. Lichens release acidic molecules that slowly dissolve the rock surface. These processes are very slow.

Lichens can also participate in stabilising soil surfaces. This occurs when communities of soil organisms including lichens, algae and other microorganisms form a crust on the soil surface that protects the soil from water and wind erosion.

Once soils have formed, microorganisms continue to play important roles in binding soil particles. Hyphae create a loose weft that aggregates larger particles. Some hyphae and bacteria also produce polysaccharide gums that assist in aggregation of finer soil particles. Earthworms ingest soil, including the microorganisms within it, and additional soil binding processes occur as a result.

Organic Matter Breakdown

Heterotrophic microorganisms, which get their energy from other organic material, are the most abundant of all soil microorganisms. They are involved in the breakdown and subsequent cycling of nutrients contained in organic matter (Waid 1984). Therefore, they are most active when there is a supply of organic matter to colonise. At any one point in time, most of the organic material in the soil has already been degraded to some extent.

Once organic matter has been partially degraded it is difficult for the soil microorganisms to further degrade it. Therefore, organic matter remains for long periods in the soil despite the presence of soil organisms.

Different soil types have characteristic threshold quantities of microorganisms that they can sustain. The number of organisms returns to a threshold level. This happens because (i) the parts of the organic matter that are easily accessible physically and chemically degrade quickly, (ii) microorganisms are eaten by soil animals (e.g. nematodes and amoebae), and (iii) unprotected microorganisms may become desiccated and die.

Biodegradation of Complex Agricultural or Industrial Chemicals

Biodegradation is a familiar concept. An important example of this process involves the transformation of pesticides by heterotrophic and autotrophic organisms in the soil. During biodegradation the microorganisms may acquire energy or carbon from degrading the chemicals and they can also release complexes of enzymes into the soil. It is interesting to note that there are similarities between the enzymatic processes involved in degradation of toxic substances and the degradation of some of the complex structural components of plants, such as wood.

Transformation of Inorganic Molecules

Autotrophic microorganisms are not directly involved in the breakdown of organic material in soil. Rather, they are important in altering the chemical structure of inorganic compounds (i.e. those that do not contain carbon) from one state to another. For example, autotrophic nitrifying bacteria convert ammonium to nitrite and subsequently to nitrate. Nitrifying bacteria have extensive membrane surfaces within their cells that are important for these transformations (Brock *et al.* 1979).

Beneficial Associations with Plants

Various soil organisms form either specific or less specific associations with roots and interact with plants in their uptake of nutrients from soil. Some organisms live in loose associations on the surface of roots or in the vicinity of the roots (the rhizosphere). Others form symbiotic associations and assist in the uptake of phosphorus (e.g. arbuscular mycorrhizal fungi). Root nodule bacteria and plants form highly specific associations that lead to nitrogen fixation.

Development and Prevention of Plant Disease

An important group of soil microorganisms that attract considerable attention are those microorganisms which cause plant diseases. These organisms include bacteria, fungi and soil animals such as nematodes that can become serious plant pathogens under some conditions.

Many plant pathogens are highly specific to particular plant species but others cause disease in different plant species that may not even be closely related. In addition, soils contain organisms that are naturally able to limit the proliferation of pathogenic organisms when soil conditions are suitable for their growth and activity. This process of disease prevention of one microorganism by another is called 'biological control'.

Seepage and Stability Analysis of Embankments During Flood Events

Mechanisms of failure of artificial levees and natural riverbanks represent a common cause of inundation and their study has relevant importance in the perspective of flood risk assessment and management. Failures of artificial levees may be due to a variety of mechanisms, including overtopping, piping, mass movements, riverbank erosion (Fread, 1985; Osman & Thorne, 1988; Gilvear et al., 1994; Torrey III, 1995; Colleselli, 1996). Many of these processes are related to variations in pore water pressures during a flood event induced by rainfall and changes in river stage.

Relationships between pore water pressure conditions within an artificial levee, river stages, and rainfall during flood events, and mutual role of these factors in triggering failures are still not well understood because of difficulties associated with monitoring the parameters involved.

Some studies have focused on groundwater movement and bank storage within a natural streambank in response to river stage fluctuations (Cooper & Rorabaugh, 1963; Marino, 1975; Sharp, 1977). However few data regarding monitoring of pore water pressures in natural riverbanks (Casagli et al., 1999; Rinaldi et al., 2000; Simon et al., 2000) and/or artificial levees (Deangelis et al., 1996) during flow events are reported in literature.

The present study is part of a general research activity regarding monitoring and modelling of saturated/unsaturated flow and analysis of failure mechanisms in natural riverbanks and artificial levees, with the overall objectives to achieve a better understanding of the factors and mechanisms determining instability and to improve the existing methods of analysis.

Four years of monitoring of pore water pressures have been conducted in a natural riverbank along the Sieve River, a gravel-bed river in Tuscany (Italy). Results of this monitoring activity are partially reported in Casagli et al. (1999). A complete bank stability analysis has been subsequently performed combining finite element seepage analysis of the saturated/unsaturated flow during a flood event with the limit equilibrium method to calculate the safety factor at different temporal steps of the event (Affuso et al., 2000; Rinaldi et al., 2000).

Recently, a research project has been undertaken with the aim to monitor the possible causes of failure of an embankment (composed by a bank in reinforced soil and an artificial levee) in another reach of the Sieve R., with particular regard to the pore water pressures and their response to external hydrological factors (rainfall, river stage variations) (Canuti et al., 2001). Monitoring activity has recently started (November 2000), with an instrumentation consisting of a battery of ten tensiometres, two piezometres, a river stage gauge, a rain gauge, and a TDR (time domain reflectometry probe), all connected to a data logger.

In this study, a preliminary analysis is conducted to verify the stability of the embankment, with the specific aim to take into account the variations of pore water pressures and confining river pressures during the different phases of a given flood event.

In fact, schematisation of flow network in steady conditions is not adequate to consider the changes in river stage and pore water pressures during a flood event, therefore a seepage analysis in transient conditions is necessary to allow an appropriate stability analysis. Some attempt has been done to consider the effects of variation of river stage using a hydrograph of simple form (Supino, 1965), while more recently Deangelis et al. (1996) carried out an analysis in transient conditions using a two-dimensional mathematical model of unsteady saturated flow.

In this paper, the method of analysis adopted includes the two following steps:

(a) finite element seepage analysis in transient conditions, to model the saturated/unsaturated flow within the embankment;

(b) stability analysis, to calculate with limit equilibrium method the safety factor in different time steps of the event.

Comparison of computed data of pore water pressures with data measured during the recently started monitoring activity will aid to verify the seepage model and the results of this analysis in a following step of the research.

Case Study and Methods

A reach of unstable riverbanks along the Sieve River (Tuscany, Central Italy) has recently been stabilised by reinforced soil and reconstruction of an artificial levee.

Sieve River has a catchment area of about 840 km^2, and a total length of about 58 km. Climate is Mediterranean and mean values of yearly rainfall range from 1000 to 1400 mm.

The study reach is located in the middle portion of the basin, having a drainage area of about 470 km^2. Average bed channel slope is about 0.0022; bed sediments are predominantly composed of gravel (D_{50} of about 94 mm).

The methodology is based on coupling seepage and stability analysis in order to take in account changes of pore water pressures during a single flood event and their effects on stability. Hydrograph and rainfall of flood events with high return periods, obtained by standard hydrologic-hydraulic procedures, are used as boundary conditions of the analysis. Finite element seepage analysis is used in transient conditions, to model the saturated/unsaturated flow within the reinforced soil and the artificial levee and obtain the pore water pressure distributions. Stability analysis is then used to calculate with limit equilibrium method the safety factor in different time steps of the event. Furthermore a probabilistic analysis, using Montecarlo method, is performed to estimate the probability of bank collapse at each time step of the flood event.

Seepage Analysis

Unsaturated and saturated flow within the streambank stabilised by reinforced soil and the artificial levee was modelled using a finite element seepage analysis (SEEP/W v.4, produced by *Geo-Slope International Ltd.*).

The model is based on the equations of motion and mass conservation. The first is the law of Darcy, originally derived for saturated soil, to simulate the flow of water through both saturated and unsaturated soil.

In this case, the hydraulic conductivity is a function of water content and of pore-water pressure. Relationships between volumetric water content and pore water pressure, and between hydraulic conductivity and pore water pressure are defined by the volumetric water content function (characteristic curve) and hydraulic conductivity function (k-curve), respectively (Brooks & Corey, 1964; Fredlund & Rahardjo, 1993). The mass conservation equation is in the form extended to unsaturated conditions (Richards, 1931; Fredlund & Rahardjo, 1993).

The embankment was discretised into 1762 quadrilateral and triangular elements and divided into four regions, composed of materials with different properties.

Different volumetric water content and hydraulic conductivity functions were assigned to the four materials based on existing curves reported in literature. At this stage of the research, the behaviour of the stabilising intervention as function of various possible soil properties employed was analysed. To this aim, the three following combinations of material properties (in terms of characteristic and k-curves) were used for two of the four regions (bank in reinforced soil and artificial levee): a) more permeable case: reinforced soil and artificial levee composed by sand with k_{sat} of 1 10^{-5} m/sec; b) intermediate case: reinforced soil composed by silty sand with k_{sat} of 1.7 10^{-6} m/sec; artificial levee composed by silty sand with k_{sat} of 1.4 10^{-6} m/sec; c) less permeable case: reinforced soil and artificial levee composed by silty clay with k_{sat} of 1.5 10^{-8} m/sec.

The analysis was performed in transient conditions, using the hydrograph as boundary condition in terms of total head versus time for the nodes along the embankment profile, while a rainfall intensity versus time function was assigned to the nodes of the artificial levee not reached by the river stage.

The hydrograph with 200 years return period was divided in 23 time steps. The initial conditions were defined considering the water table at the same level of the river stage before the start of the flood event, and varying negative pore water pressures hydrostatically with distance for the unsaturated portion of the embankment, setting an upper limit at 1 m above the water table (non-zero surface flux).

Stability Analysis

The stability analysis of the embankment was performed using a software (SLOPE/W v.4, produced by *Geo-Slope International Ltd.*) for the application of limit equilibrium methods. Only possible mass failures of the embankment or portions of it are considered, not including failures by overtopping or piping.

The safety factor was computed with the Morgestern-Price method for each time step of the flood event adopted in the seepage analysis, considering the surface failure with the minimum value. Pore-water pressures at the base of each slide have been specified using the pore water pressure distribution obtained by the seepage analysis for each time step of the computation.

The Mohr-Coulomb criterion in terms of effective stress was used in case of positive pore pressures and the Fredlund et al. (1978) criterion for the portions of the embankment with negative pore pressures. Different shear strength parameters, according to typical values proposed in literature, were assigned to the various materials used for the three scenarios (more permeable, intermediate, less permeable) (c' ranging from 0.5 to 2 kPa; f ranging from 25° to 30°; f^b ranging from 10° to 20°).

Effects of geosynthetics reinforcement sheets into the portion of reinforced soil were taken into account as anchor loads in the stability analysis, specifying embedment lengths and the tear strength of the fabric. Only the overall stability analysis of the reinforced soils was conducted, not considering possible local failures (direct sliding or bond: Jewell, 1990) internal to the reinforced slope.

A probabilistic stability analysis using Monte Carlo method was also performed to consider the variability of the input parameters and to express the stability conditions in terms of probability of failure. The variability of soil properties and shear strength parameters (bulk unit weight *g*, *c'*, *f*, *f* ^{b}) of the different materials was considered, and assumed to be normally distributed.

The probabilistic analysis was again conducted for each time step in which the flood event was divided, having as result the probability of failure of the embankment, corresponding to the probability of obtaining a factor of safety less than 1.0 and computed by integrating the area under the probability density function for factors of safety less than 1.0.

Results And Discussion

Pore water pressure distribution for the time step 19 of the seepage analysis, corresponding to the instant of the event (about ten hours after the peak flood) when the safety factor was minimum. Time variation of piezometric surface relative to the 23 time steps of seepage simulation, illustrating the portion of the embankment completely saturated during the event. Results are referred to the more permeable case, which appears to be the more realistic scenario as confirmed from field data recently collected (Canuti et al., 2001).

Time variations of the safety factor and the probability of failure compared with the river stage hydrograph for the simulated event. It can be noted how both the parameters related to the stability conditions of the embankment show a significant variation during the flood event.

This is due to the effects of both changes of the pore water pressures within the embankment, as result of lateral and vertical infiltration of water, and of the confining pressures of the river. During the rising phase of the hydrograph, pore water pressures progressively increase, inducing a decrease in effective stresses due to development of positive pore water pressures in the lower portion of the slope and a decrease of stabilising suction forces in the unsaturated upper region. However, safety factor tends progressively to increase because of the prevailing effect of increasing confining pressures of the river.

Probabilities of failure tend to be zero during this phase. During the drawdown phase of the flood event, the two previous trends are reverted. Destabilising positive pore water pressures tend progressively to dissipate and stabilising confining river pressures decrease; however, the rate of river stage lowering is higher than the rate of phreatic surface lowering, due to the relatively low velocities of the water within the soil.

This difference of processes leads the safety factor to lower values than the initial one, reaching a minimum with a delay of about 10 hours from the peak stage; at the same time the probability of failure is maximum. Subsequently, the safety factor tends slowly towards the initial value as dissipation of pore water pressures continues, requiring a time that is function of the soil properties and the hydrograph characteristics.

It has to be remarked how the instability conditions during a flood event are strongly related to the initial pore water pressure conditions, affected by rainfall and/or minor flow events preceeding a flood event with high return period. This aspect is obviously not taken in consideration in the present analysis, even though it can be easily included by simulating a succession of various flow events.

Stability of an embankment has been analysed in non-steady flow simulating seepage processes in saturated and unsaturated conditions during a flood event in a river reach of Central Italy. In particular, a hydrograph and a rainfall intensity of 200 years return period have been used as boundary conditions of the analysis. The embankment is composed by a lower bank of reinforced soil and an artificial levee in the upper part of silty-sand material.

The event duration, of about 45 hours, has been discretized in 23 time steps of variable length ranging from half to two hours. Finite element seepage model has been coupled with limit equilibrium method for stability analysis at different time steps of the flood event. A probabilistic stability analysis (Monte Carlo method) was also carried out to account for the uncertainties in soil strength parameters and to express the stability conditions in terms of probability of failure.

Results have shown the influence of soil properties, characteristics of the hydrograph, and pore water pressures on the general stability of the embankment.

In particular, stability conditions during a flood event varies through time due to rapid changes in pore water pressures and confining river pressures. During the rising phase of the hydrograph, safety factor increases as a result of the stabilising confining pressure of the water in the river although pore water pressures are also increasing in this phase.

Safety factor and probability of failure reach critical conditions during the drawdown phase of the hydrograph, about ten hours after the peak stage, when a partial loss of confining pressure is combined with relatively high pore water pressures.

For this reason, the more common approach based on a steady flow analysis does not appear suitable to take in account the complex changes in pore water and confining river pressures and their effects on stability, leading to an unappropriate estimation of the safety factor. Therefore, the present approach seems to be suitable in investigating stability of artificial levees and their probabilities of failure for a given geometry, soil properties and hydrograph characteristics, with possible implications for flood risk assessment and management. The methodology is also adequate to design artificial levees, allowing appropriate selection of geometry and soil characteristics to be employed for a river reach with given flood hydrograph.

Soil Thermal Properties

The thermal properties of soil are a component of soil physics that has found important uses in engineering, climatology and agriculture. These properties influence how energy is partitioned in the soil profile. While related to soil temperature, it is more accurately associated with the transfer of heat throughout the soil, by radiation, conduction and convection.

The main soil thermal properties are:

- Volumetric heat capacity, SI Units: $J.m^{-3}.K^{-1}$
- Thermal conductivity, SI Units: $W.m^{-1}.K^{-1}$
- Thermal diffusivity, SI Units: $m^2.s^{-1}$

Measurement

It is hard to say something general about the soil thermal properties at a certain location because these are in a constant state of flux from diurnal and seasonal variations. Apart from the basic soil composition, which is constant at one location, soil thermal properties are strongly influenced by the soil volumetric water content, volume fraction of solids and volume fraction of air. Air is a poor thermal conductor and reduces the effectiveness of the solid and liquid phases to conduct heat. While the solid phase has the highest conductivity it is the variability of soil moisture that largely determines thermal conductivity.

As such soil moisture properties and soil thermal properties are very closely linked and are often measured and reported together. Temperature variations are most extreme at the surface of the soil and these variations are transferred to sub surface layers but at reduced rates as depth increases. Additionally there is a time delay as to when maximum and minimum temperatures are achieved at increasing soil depth (sometimes referred to as thermal lag).

One possible way of assessing soil thermal properties is the analysis of soil temperature variations versus depth Fourier's law,

$$Q = -\lambda dT / dz$$

where Q is heat flux or rate of heat transfer per unit area J m^{-2}.s^{-1} or W m^{-2}, ë is thermal conductivity W m^{-1}.K^{-1}; dT/dz is the gradient of temperature (change in temp/change in depth) K m^{-1}.

The most commonly applied method for measurement of soil thermal properties, is to perform in-situ measurements, using Non-Steady-State Probe systems, or Heat Probes.

Single and Dual Heat Probes

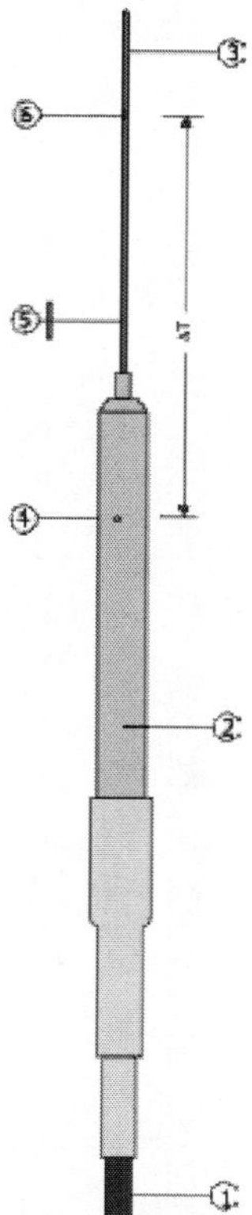

Figure: *Small Size Non-Steady-State Probe: The probe consists of a needle (3) with a single thermocouple junction (6) and a heating wire, (5). It is inserted into the medium that is investigated.*

The single probe method employs a heat source inserted into the soil whereby heat energy is applied continuously at a given rate. The thermal properties of the soil can be determined by analysing the temperature response adjacent to the heat source via a thermal sensor. This method reflects the rate at which heat is conducted away from the probe. The limitation of this device is that it measures thermal conductivity only. Applicable standards are: IEEE Guide for Soil Thermal Resistivity Measurements (IEEE Standard 442-1981) as well as with ASTM D 5334-08 Standard Test Method for Determination of Thermal Conductivity of Soil and Soft Rock by Thermal Needle Probe Procedure.

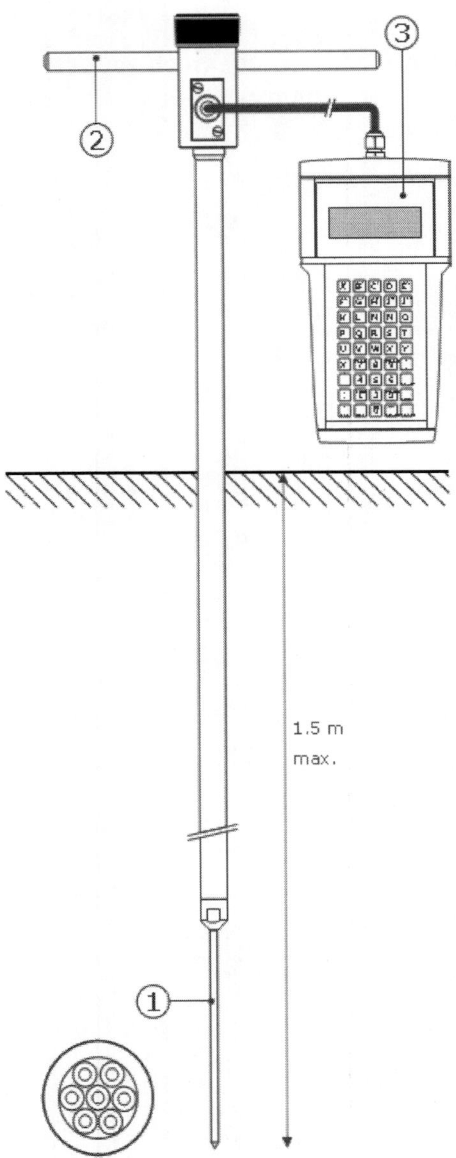

***Figure:** Example of a complete system for measurement of soil thermal conductivity, specially designed for measurements at around 1.5 metres below the soil surface, which is the typical depth of burial for high voltage cables.*

After further research the dual-probe heat-pulse technique was developed. It consists of two parallel needle probes separated by a distance (r). One probe contains a heater and the other a temperature sensor. The dual probe device is inserted into the soil and a heat pulse is applied and the temperature sensor records the response as a function of time. That is, a heat pulse is sent from the probe across the soil (r) to the sensor.

The great benefit of this device is that it measures both thermal diffusivity and volumetric heat capacity. From this, thermal conductivity can be calculated meaning the dual probe can determine all the main soil thermal properties. Potential drawbacks of the heat-pulse technique have been noted.

Remote Sensing

Remote sensing from satellites, aircraft has greatly enhanced how the variation in soil thermal properties can be identified and utilised to benefit many aspects of human endeavour. While remote sensing of reflected light from surfaces does indicate thermal response of the topmost layers of soil (a few molecular layers thick), it is thermal infrared wavelength that provides energy variations extending to varying shallow depths below the ground surface which is of most interest. A thermal sensor can detect variations to heat transfers into and out of near surface layers because of external heating by the thermal processes of conduction, convection, and radiation. Microwave remote sensing from satellites has also proven useful as it has an advantage over TIR of not being effected by cloud cover.

The various methods of measuring soil thermal properties have been utilised to assist in diverse fields such as; the expansion and contraction of construction materials especially in freezing soils, longevity and efficiency of gas pipes or electrical cables buried in the ground, energy conservation schemes, in agriculture for timing of planting to ensure optimum seedling emergence and crop growth, measuring greenhouse gas emissions as heat effects the liberation of carbon dioxide from soil. Soil thermal properties are also becoming important in areas of environmental science such as determining water movement in radioactive waste and in locating buried land mines.

Permeability (Earth Sciences)

Permeability in fluid mechanics and the earth sciences (commonly symbolised as κ, or k) is a measure of the ability of a porous material (often, a rock or unconsolidated material) to allow fluids to pass through it.

Units

The SI unit for permeability is m^2. A traditional unit for permeability is the *darcy* (D), or more commonly the *millidarcy* (mD) (1 darcy ≈ $10^{-12} m^2$). The unit of cm^2 is also sometimes used ($1\ m^2 = 10^4\ cm^2$).

Applications

The concept of permeability is of importance in determining the flow characteristics of hydrocarbons in oil and gas reservoirs, and of groundwater in aquifers.

For a rock to be considered as an exploitable hydrocarbon reservoir without stimulation, its permeability must be greater than approximately 100 mD (depending on the nature of the hydrocarbon - gas reservoirs with lower permeabilities are still exploitable because of the lower viscosity of gas with respect to oil). Rocks with permeabilities significantly lower than 100 mD can form efficient *seals*. Unconsolidated sands may have permeabilities of over 5000 mD.

The concept has also many practical applications outside of geology, for example in chemical engineering (e.g., filtration).

Description

Permeability is part of the proportionality constant in Darcy's law which relates discharge (flow rate) and fluid physical properties (e.g. viscosity), to a pressure gradient applied to the porous media:

$$v = \frac{\kappa}{\mu}\frac{\Delta P}{\Delta x}$$

Therefore:

$$\kappa = v\frac{\mu \Delta x}{\Delta P}$$

where:

v is the superficial fluid flow velocity through the medium (i.e., the average velocity calculated as if the fluid were the only phase present in the porous medium) (m/s)

κ is the permeability of a medium (m^2)

μ is the dynamic viscosity of the fluid (Pa ·s)

ΔP is the applied pressure difference (Pa)

Δx is the thickness of the bed of the porous medium (m)

In naturally occurring materials, permeability values range over many orders of magnitude.

Relation to Hydraulic Conductivity

The proportionality constant specifically for the flow of water through a porous media is called the hydraulic conductivity; permeability is a portion of this, and is a property of the porous media only, not the fluid. Given the value of hydraulic conductivity for a subsurface system, *k*, the permeability can be calculated as:

$$\kappa = k\frac{\mu}{\rho g}$$

where

- κ is the permeability, m^2
- k is the hydraulic conductivity, m/s
- μ is the dynamic viscosity, kg/(m ·s)
- ρ is the density of the fluid, kg/m^3
- g is the acceleration due to gravity, m/s^2.

Determination

Permeability is typically determined in the lab by application of Darcy's law under steady state conditions or, more generally, by application of various solutions to the diffusion equation for unsteady flow conditions.

Permeability needs to be measured, either directly (using Darcy's law) or through estimation using empirically derived formulas. However, for some simple models of porous media, permeability can be calculated (e.g., random close packing of identical spheres).

Permeability Model Based on Conduit Flow

Based on Hagen–Poiseuille equation for viscous flow in a pipe, permeability can be expressed as:

$$\kappa_I = C \cdot d^2$$

where:

κ_I is the intrinsic permeability [length^2]

C is a dimensionless constant that is related to the configuration of the flow-paths

d is the average, or effective pore diametre [length].

Intrinsic and Absolute Permeability

The terms *intrinsic permeability* and *absolute permeability* states that the permeability value in question is an intensive property (not a spatial average of a heterogeneous block of material), that it is a function of the material structure only (and not of the fluid), and explicitly distinguishes the value from that of relative permeability.

Permeability to Gases

Sometimes permeability to gases can be somewhat different that those for liquids in the same media. One difference is attributable to "slippage" of gas at the interface with the solid when the gas mean free path is comparable to the pore size (about 0.01 to 0.1 μm at standard temperature and pressure). For example, measurement of permeability through sandstones and shales yielded values from 9.0×10^{-19} m^2 to 2.4×10^{-12} m^2 for water and between 1.7×10^{-17} m^2 to 2.6×10^{-12} m^2 for nitrogen gas.

Tensor Permeability

To model permeability in anisotropic media, a permeability tensor is needed. Pressure can be applied in three directions, and for each direction, permeability can be measured (via Darcy's law in 3D) in three directions, thus leading to a 3 by 3 tensor. The tensor is realised using a 3 by 3 matrix being both symmetric and positive definite (SPD matrix):

- The tensor is symmetric by the Onsager reciprocal relations.
- The tensor is positive definite as the component of the flow parallel to the pressure drop is always in the same direction as the pressure drop.

The permeability tensor is always diagonalizable (being both symmetric and positive definite). The eigenvectors will yield the principal directions of flow, meaning the directions where flow is parallel to the pressure drop, and the eigenvalues representing the principal permeabilities.

Relative Permeability

In multiphase flow in porous media, the relative permeability of a phase is a dimensionless measure of the effective permeability of that phase. It is the ratio of the effective permeability of that phase to the absolute permeability. It can be viewed as an adaptation of Darcy's law to multiphase flow.

For two-phase flow in porous media given steady-state conditions, we can write

$$q_i = -\frac{\kappa_i}{\mu_i}\nabla P_i \qquad \text{for} \quad i = 1,2$$

where q_i is the flux, ∇P_i is the pressure drop, μ_i is the viscosity. The subscript *i* indicates that the parameters are for phase *i*.

κ_i is here the phase permeability (i.e., the effective permeability of phase *i*), as observed through the equation above.

Relative permeability, κ_{ri}, for phase *i* is then defined from $\kappa_i = \kappa_{ri}\kappa$ as

$$\kappa_{ri} = \kappa_i / \kappa$$

where κ is the permeability of the porous medium in single-phase flow, i.e., the absolute permeability. Relative permeability must be between zero and one.

In applications, relative permeability is often represented as a function of water saturation however due to capillary hysteresis, one often resorts to one function or curve measured under drainage and one measured under imbibition.

Under this approach, the flow of each phase is inhibited by the presence of the other phases. Thus the sum of relative permeabilities over all phases is less than 1.

However, apparent relative permeabilities larger than 1 have been obtained since the Darcean approach disregards the viscous coupling effects derived from momentum transfer between the phases. This coupling could enhance the flow instead of inhibit it. This has been observed in heavy oil petroleum reservoirs when the gas phase flows as bubbles or patches (disconnected).

Assumptions

The above form for Darcy's law is sometimes also called Darcy's extended law, formulated for horizontal, one-dimensional, immiscible multiphase flow in homogeneous and isotropic porous media. The interactions between the fluids are neglected, so this model assumes that the solid porous media and the other fluids form a new porous matrix through which a phase can flow, implying that the fluid-fluid interfaces remain static in steady-state flow, which is not true, but this approximation has proven useful anyway.

Each of the phase saturation must be larger than the irreducible saturation, and each phase is assumed continuous within the porous medium.

Approximations

Based on experimental data, simplified models of relative permeability as a function of water saturation can be constructed.

Corey-type

An often used approximation of relative permeability is the Corey correlation which is power law in the water saturation S_w. If S_{wi} (also denoted S_{wir}, or S_{wr}, or S_{wc}) is the irreducible (minimal) water saturation, and S_{orw} is the residual (minimal) oil saturation after water flooding, we can define a normalised (or scaled) water saturation value

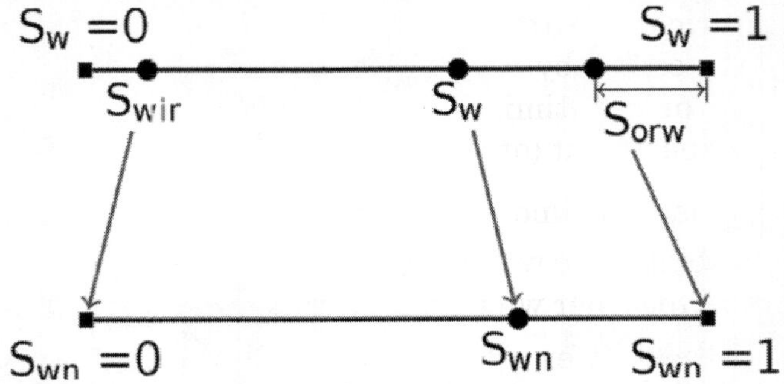

***Figure:** Normalisation of water saturation values.*

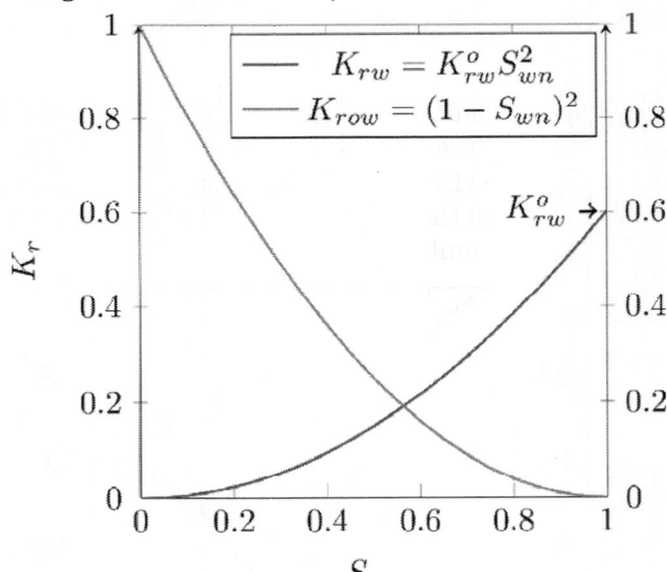

***Figure:** Example of Corey approximation in normalised* S_w *coordinates, here* $N_o = N_w = 2$.

$$S_{wn}(S_w) = \frac{S_w - S_{wi}}{1 - S_{wi} - S_{orw}}$$

The Corey correlations of the relative permeability of oil and water are then

$$K_{row}(S_w) = (1 - S_{wn}(S_w))^{N_o} \text{ and}$$

$$K_{rw}(S_w) = K^o_{rw} S_{wn}(S_w) N_w$$

when the permeability basis is oil with irreducible water present.

We note the desired properties.

$$K_{row}(S_{wi}) \;=1 \quad K_{row}(1-S_{orw}) \;=0$$
$$K_{rw}(S_{wi}) \;=0 \quad K_{rw}(1-S_{orw}) \;=K^o_{rw}$$

The empirical parameters N_o and N_w can be obtained from measured data either by optimising to analytical interpretation of measured data, or by optimising using a core flow numerical simulator to match the experiment (often called history matching).

$N_o = N_w = 2$ is sometimes appropriate. The physical property K^o_{rw} is called the end point of the water relative permeability, and it is obtained either before or together with the optimising of N_o and N_w.

In case of gas-water system or gas-oil system there are Corey correlations similar to the oil-water relative permeabilities correlations shown above.

LET-type

The Corey approximation only has one degree of freedom for the oil relative permeability and two degrees of freedom for the water permeability (in S_{wn}). The LET-correlation adds more degrees of freedom in order to accommodate the shape of measured relative permeability curves in SCAL experiments.

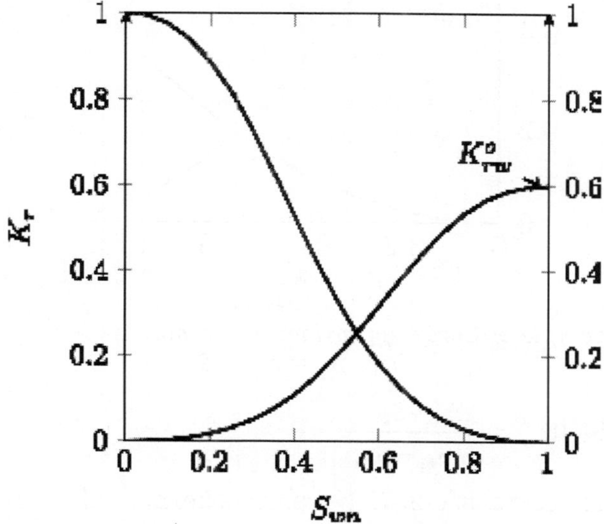

Figure: *Example of LET-correlation with L,E,T all equal to 2, and* $K^o_{rw} = 0.6$. *Normalised* S_w *coordinates.*

The LET-type approximation is described by 3 parameters L, E, T. The correlation for water and oil relative permeability with water injection is thus

$$K_{rw} = \frac{K^{o}_{rw} S_{wn}^{L_w}}{S_{wn}^{L_w} + E_w (1 - S_{wn})^{T_w}}$$

and

$$K_{row} = \frac{(1 - S_{wn})^{L_o}}{(1 - S_{wn})^{L_o} + E_o S_{wn}^{T_o}}$$

written using the same S_w normalisation as for Corey.

Only S_{wi}, S_{orw} and K^{o}_{rw} have direct physical meaning, while the parameters *L*, *E* and *T* are empirical. The parameter *L* describes the lower part of the curve, and by similarity and experience the *L*-values are comparable to the appropriate Corey parameter.

The parameter *T* describes the upper part (or the top part) of the curve in a similar way that the *L*-parameter describes the lower part of the curve. The parameter *E* describes the position of the slope (or the elevation) of the curve.

A value of one is a neutral value, and the position of the slope is governed by the *L*- and *T*-parameters. Increasing the value of the *E*-parameter pushes the slope towards the high end of the curve. Decreasing the value of the *E*-parameter pushes the slope towards the lower end of the curve. Experience using the LET correlation indicates that the parameter, *L* e^{-} 0.3, $E > 0.3$ and *T* e^{-} 0.3.

In case of gas-water system or gas-oil system there are LET correlations similar to the oil-water relative permeabilities correlations shown above.

Hydraulic Conductivity

Hydraulic conductivity, symbolically represented as *K*, is a property of vascular plants, soil or rock, that describes the ease with which water can move through pore spaces or fractures.

It depends on the intrinsic permeability of the material and on the degree of saturation. Saturated hydraulic conductivity, K_{sat}, describes water movement through saturated media.

Methods of Determination

There are two broad categories of determining hydraulic conductivity:

- *Empirical* approach by which the hydraulic conductivity is correlated to soil properties like pore size and particle size (grain size) distributions, and soil texture
- *Experimental* approach by which the hydraulic conductivity is determined from hydraulic experiments using Darcy's law

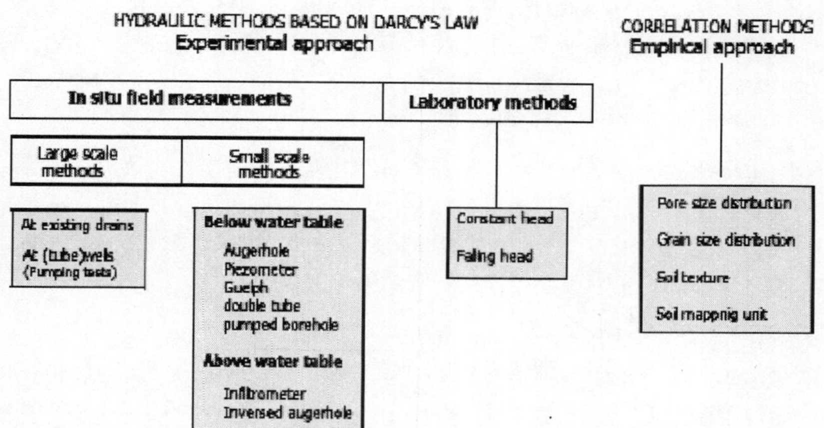

Figure: *Overview of determination methods*

The experimental approach is broadly classified into:

- Laboratory tests using soil samples subjected to hydraulic experiments
- *Field tests* (on site, in situ) that are differentiated into:
    - o small scale field tests, using observations of the water level in cavities in the soil
    - o large scale field tests, like pump tests in wells or by observing the functioning of existing horizontal drainage systems.

The small scale field tests are further subdivided into:

- infiltration tests in cavities *above* the water table
- slug tests in cavities *below* the water table

Estimation by Empirical Approach

Estimation from Grain Size

Shepherd derived an empirical formula for approximating hydraulic conductivity from grain size analyses:

$$K = a(D_{10})^b$$

where

a and *b* are empirically derived terms based on the soil type, and

D_{10} is the diametre of the 10 percentile grain size of the material

Pedotransfer Function

A pedotransfer function (PTF) is a specialised empirical estimation method, used primarily in the soil sciences, however has increasing use in hydrogeology. There are many different PTF methods, however, they all attempt to determine soil properties, such as hydraulic conductivity, given several measured soil properties, such as soil particle size, and bulk density.

Determination by Experimental Approach

There are relatively simple and inexpensive laboratory tests that may be run to determine the hydraulic conductivity of a soil: constant-head method and falling-head method.

Laboratory Methods

Constant-Head Method: The constant-head method is typically used on granular soil. This procedure allows water to move through the soil under a steady state head condition while the quantity (volume) of water flowing through the soil specimen is measured over a period of time.

By knowing the quantity Q of water measured, length L of specimen, cross-sectional area A of the specimen, time t required for the quantity of water Q to be discharged, and head h, the hydraulic conductivity can be calculated:

$$Q = Av$$

where v is the flow velocity. Using Darcy's Law:

$$v = Ki$$

and expressing the hydraulic gradient i as:

$$i = \frac{h}{L}$$

where h is the difference of hydraulic head over distance L, yields:

$$Q = \frac{AKh}{L}$$

Solving for K gives:

$$K = \frac{QL}{Ah}$$

Falling-Head Method

The falling-head method is totally different than the constant head methods in its initial setup; however, the advantage to the falling-head method is that can be used for both fine-grained and coarse-grained soils. The soil sample is first saturated under a specific head condition. The water is then allowed to flow through the soil without maintaining a constant pressure head.

$$K = \frac{2.3aL}{At}\log\left(\frac{h_1}{h_2}\right)$$

In-situ (Field) Methods

Augerhole Method: There are also in-situ methods for measuring the hydraulic conductivity in the field.

When the water table is shallow, the augerhole method, a slug test, can be used for determining the hydraulic conductivity below the water table.

The method was developed by Hooghoudt (1934) in The Netherlands and introduced in the US by Van Bavel en Kirkham (1948).

The method uses the following steps:

1. an augerhole is perforated into the soil to below the water table
2. water is bailed out from the augerhole
3. the rate of rise of the water level in the hole is recorded
4. the K-value is calculated from the data as :

$$K_h = C\ (Ho\text{-}Ht)\ /\ t$$

where: K_h = horizontal saturated hydraulic conductivity (m/day), H = depth of the waterlevel in the hole relative to the water table in the soil (cm), Ht = H at time t, Ho = H at time t = 0, t = time (in seconds) since the first measurement of H as Ho, and F is a factor depending on the geometry of the hole:

$$F = 4000r\ /\ h'(20+D/r)(2''h'/D)$$

where: r = radius of the cylindrical hole (cm), h' is the average depth of the water level in the hole relative to the water table in the soil (cm), found as h'=(Ho+Ht)/2, and D is the depth of the bottom of the hole relative to the water table in the soil (cm).

The picture shows a large variation of K-values measured with the augerhole method in an area of 100 ha. The ratio between the highest and lowest values is 25. The cumulative frequency distribution is lognormal and was made with the CumFreq program.

Related Magnitudes

Transmissivity: An aquifer may consist of n soil layers. The *transmissivity* for horizontal flow (T_i) of the i " *th* soil layer with a *saturated* thickness d_i and horizontal hydraulic conductivity Kh_i is:

$$T_i = Kh_i\, d_i$$

Transmissivity is directly proportional to horizontal hydraulic conductivity (Kh_i) and thickness (d_i). Expressing Kh_i in m/day and d_i in m, the transmissivity (T_i) is found in units m^2/day.

The transmissivity is a measure of how much water can be transmitted horizontally, such as to a pumping well.

Transmissivity should not be confused with the similar word transmittance used in optics, meaning the fraction of incident light that passes through a sample.

The total transmissivity (Tt) of the aquifer is :

$$Tt = \Sigma\, T_i = \Sigma\, Kh_i\, d_i$$

where Σ signifies the summation over all layers: i= 1, 2, 3, . . .n.

The *apparent* horizontal hydraulic conductivity (Kh_A) of the aquifer is:

$$Kh_A = Tt / Dt$$

where Dt is the total thickness of the aquifer: Dt= $\Sigma\, d_i$, with i= 1, 2, 3, . . .n

The transmissivity of an aquifer can be determined from pumping tests.

When a soil layer is above the water table, it is not saturated and does not contribute to the transmissivity. When the soil layer is entirely below the water table, its saturated thickness corresponds to the thickness of the soil layer itself. When the water table is inside a soil layer, the saturated thickness corresponds to the distance of the water table to the bottom of the layer. As the water table may behave dynamically, this thickness may change from place to place or from time to time, so that the transmissivity may vary accordingly.

In a semi-confined aquifer, the water table is found within a soil layer with a negligibly small transmissivity, so that changes of the total transmissivity (Dt) resulting from changes in the level of the water table are negligibly small. When pumping water from an unconfined aquifer, where the water table is inside a soil layer with a significant transmissivity, the water table may be drawn down whereby the transmissivity reduces and the flow of water to the well diminishes.

Resistance

The *resistance* to vertical flow (R_i) of the *i* " *th* soil layer with a *saturated* thickness d_i and vertical hydraulic conductivity Kv_i is:

$$R_i = d_i / Kv_i$$

Expressing Kv_i in m/day and d_i in m, the resistance (R_i) is expressed in days.

The total resistance (Rt) of the aquifer is :

$$Rt = \Sigma R_i = \Sigma d_i / Kv_i$$

where Σ signifies the summation over all layers: *i*= 1, 2, 3, . . .*n*

The *apparent* vertical hydraulic conductivity (Kv_A) of the aquifer is:

$$Kv_A = Dt / Rt$$

where Dt is the total thickness of the aquifer: $Dt = \Sigma d_i$, with *i*= 1, 2, 3, . . .*n*

The resistance plays a role in aquifers where a sequence of layers occurs with varying horizontal permeability so that horizontal flow is found mainly in the layers with high horizontal permeability while the layers with low horizontal permeability transmit the water mainly in a vertical sense.

Anisotropy

When the horizontal and vertical hydraulic conductivity (Kh_i and Kv_i) of the *i* " *th* soil layer differ considerably, the layer is said to be anisotropic with respect to hydraulic conductivity.

When the *apparent* horizontal and vertical hydraulic conductivity (Kh_A and Kv_A) differ considerably, the aquifer is said to be anisotropic with respect to hydraulic conductivity.

An aquifer is called *semi-confined* when a saturated layer with a relatively small horizontal hydraulic conductivity (the semi-confining layer or aquitard) overlies a layer with a relatively high horizontal hydraulic conductivity so that the flow of groundwater in the first layer is mainly vertical and in the second layer mainly horizontal.

The resistance of a semi-confining toplayer of an aquifer can be determined from pumping tests.

When calculating flow to drains or to a well field in an aquifer with the aim to control the water table, the anisotropy is to be taken into account, otherwise the result may be erroneous.

Relative Properties

Because of their high porosity and permeability, sand and gravel aquifers have higher hydraulic conductivity than clay or unfractured granite aquifers. Sand or gravel aquifers would thus be easier to extract water from (e.g., using a pumping well) because of their high transmissivity, compared to clay or unfractured bedrock aquifers.

Hydraulic conductivity has units with dimensions of length per time (e.g., m/s, ft/day and (gal/day)/ft^2); transmissivity then has units with dimensions of length squared per time. The following table gives some typical ranges (illustrating the many orders of magnitude which are likely) for *K* values.

Hydraulic conductivity (*K*) is one of the most complex and important of the properties of aquifers in hydrogeology as the values found in nature:

- range over many orders of magnitude (the distribution is often considered to be lognormal),
- vary a large amount through space (sometimes considered to be randomly spatially distributed, or stochastic in nature),
- are directional (in general *K* is a symmetric second-rank tensor; e.g., vertical *K* values can be several orders of magnitude smaller than horizontal *K* values),
- are scale dependent (testing a m^3 of aquifer will generally produce different results than a similar test on only a cm^3 sample of the same aquifer),
- must be determined indirectly through field pumping tests, laboratory column flow tests or inverse computer simulation, (sometimes also from grain size analyses), and
- are very dependent (in a non-linear way) on the water content, which makes solving the unsaturated flow equation difficult. In fact, the variably saturated *K* for a single material varies over a wider range than the saturated *K* values for all types of materials.

Slope Stability Radar

The Slope Stability Radar (SSR) is a new application for the monitoring of slope stability at open-cut mines. It is a system currently in use across the global mining and civil industries.

Slope stability is a critical safety and production issue for open-cut mines around the world to understand the probability of mine wall failures. Many fatalities have occurred in mining due to slope failure - sudden rock and wall collapses. Even when there is no injury, there is a high cost due to lost production and often damaged equipment. A common technique to determine slope stability is to monitor the small precursory movements, which occur prior to collapse.

Slope Stability Radar, as a contemporary slope stability analysis tool, have been developed to remotely scan a rock slope to monitor the spatial deformation of the face. Small movements of a rough wall can be detected with sub-millimetre accuracy by using interferometry techniques. The effects of atmospheric variations and spurious signals can be reduced via signal processing means.

The advantage of the slope stability radar over other monitoring techniques is that it provides full area coverage without the need for mounted reflectors or equipment on the wall. In addition, the radar waves adequately penetrate through rain, dust and smoke to give reliable measurements, in real-time and twenty-four hours a day.

A series of time measurements over time can be used to track movement of the rack slopes, and also detect variations in the rate of movement, which indicate instability. The radar information can be combined with various other information on a graphical display. Typical examples are 3-D models of the rock surface or pictures of the slope being monitored.

Slope stability radars are currently manufactured by companies in Australia, South Africa and Italy.

The Permeability of Soil

Rain, Rain, Will It Go Away?: Where does rainwater go? It may flow into streams and rivers, gutters and sewers, form puddles, or soak into the ground. Although rock, sand, and soil are solid, there are spaces between the grains of the material called pores.

Water may flow into these pores. The measure of how much open space there is in a solid is called its "porosity." (Look at the Absorbency of Rock experiment to learn more about porosity.)

Another important measure is permeability, which is the rate at which fluid can flow through the pores of a solid. If soil has high permeability, rainwater will soak into it easily. If the permeability is low, rainwater will tend to accumulate on the surface or flow across the surface if it is not level.

In this experiment you will be measuring the permeability of soil. You will put a bottomless metal can into soil and see how quickly the water you pour into the can flows into the ground below it.

For comparison purposes you can try more than one area or type of soil. You can also share you results with others who have tried this experiment to compare permeability rates of different soil types in different parts of the world.

Tools and Materials

You will need:

- a fruit or vegetable juice can, 1.5 litres or larger, with both ends removed
- a hammer
- a wooden board
- a ruler
- a bucket, jar, or bottle to hold 1 to 2 litres of water
- a watch
- a 10 cm long piece of masking tape or electrical tape
- pencil and paper or a computer to record your observations and result

Procedures

Here's what to do:

1. Before you disturb the soil in any way, describe it as best you can. Think about the location (pasture, riverbank, beach, etc.), the plant material that may be present (grass, moss, dead leaves, etc.) and the soil condition (dry/moist, sandy, loose granular, hard clay, etc.). Write down your observations.
2. Set the can on the ground and place the piece of wood on top. By hitting the board with the hammer, drive the can about 5 cm into the ground.
3. Put the piece of tape on the inside of the can near the top so that it is parallel with the top edge.
4. Measure the distance for the bottom of the tape to the ground and write this down.
5. Pour water into the can until it reaches the level of the bottom of the tape inside the can. Record the time.
6. As the water permeates through the soil, the water level will drop. You can determine how many centimetres of water are permeating through the soil by measuring the distance between the starting height marker and the surface of the water. Using a ruler, measure this distance 30 minutes and 60 minutes from the time you first poured the water into the can.

7. Record your measurements in the data table.
8. (If the water runs out during the course of the experiment, immediately fill it up to the tape marker again. Measurements you make after that should be recorded as the full the distance from the ground to the tape, plus the distance from the water level to the tape. If you have to fill the can again, make sure to add the distance from the ground to the tape to your measurements again.)
9. Divide the amount of water absorbed in one hour by 60 to get the permeability in centimetres per minute for the entire hour.
10. Divide the amount of water absorbed in 30 minutes by 30 to get the permeability in centimetres per minute for the first half hour. Is this the same rate as for the entire hour?

Coefficient of Permeability

Ability of porous body (soils, rocks) to transport water of given properties (e.g. ground water) is denoted as seepage. The amount of water flowing through a certain area can be represented by the coefficient of permeability.

The coefficient of permeability represents the slope of a linear dependence of water flow velocity on the hydraulic gradient (gradient of hydraulic head) in Dracy's law written as:

$$v = n.v_s = -K_r K_{sat} \nabla h$$

where:			
	v_s	-	velocity of water flowing through pores
	n	-	porosity
	K_r	-	relative coefficient of permeability
	K_{sat}	-	Permeability matrix storing coefficients of permeability of fully saturated soil $k_{x,}$ $k_{y,}$ which may be different along individual coordinate axes
	h	-	hydraulic head

Hydraulic head at a given point of region of flow is defined a sum of vertical coordinate and pressure head and as such it determines the height of water in piezometre at a given point:

$$h = \frac{p}{\gamma_w} = y$$

where: γ_w-the weight of water

Example values of coefficients of permeability for various soils (Myslivec)

Type of soil	*Coefficient of permeability k [m/day]*	*Motion of water particle by 1 cm for hydraulic gradient i=1 per time*
Soft sand	10^2 - 10	6 s – 10 min
Clayey sand	10^{-1} - 10^{-2}	100 min – 18 hrs
Loess loam	10^{-2} - 10^{-4}	18 hrs – 70 days
Loam	10^{-4} - 10^{-5}	70 days – 2 years
Clayey soil	10^{-5} - 10^{-6}	2 years – 20 years
Clay	10^{-6} - 10^{-7}	20 -200 years

There are several ways for determining the coefficient of permeability k. They grouped as follows:

a) *Laboratory measurements* : Several types are available for the range of k 10^4 - 10^{-6} m/day.

b) *Field measurements* : Dwell or sink tests, measurement of filtration velocity of flow, for the range of k 10^6 - 1 m/day.

Using Empirical Expressions

$$k = 100.d_{10}^2.e^2$$

Suitable for non-cohesive soils, k 10^6 - 10 m/day, they produce only guidance values – e.g. according Terzaghi:

---

where: d_{10} - diametre of effective solid particle

e - void ratio

By Calculation from Time Dependent Consolidation Process

One must know the coefficient of consolidation c_v and consolidation curve (semi-logarithmic dependence of deformation on time). This is only an indirect determination from the expression:

$$k = \frac{C_v.\rho_w.g.\alpha_v}{1+e_0}$$

---

where: e_0 - initial void ratio

c_v - coefficient of consolidation

ρ_w - bulk density of water

g - gravitational acceleration

a_v - coefficient of compressibility

Bearing Capacity

In geotechnical engineering, bearing capacity is the capacity of soil to support the loads applied to the ground. The bearing capacity of soil is the maximum average contact pressure between the foundation and the soil which should not produce shear failure in the soil. *Ultimate bearing capacity* is the theoretical maximum pressure which can be supported without failure; *allowable bearing capacity* is the ultimate bearing capacity divided by a factor of safety. Sometimes, on soft soil sites, large settlements may occur under loaded foundations without actual shear failure occurring; in such cases, the allowable bearing capacity is based on the maximum allowable settlement.

There are three modes of failure that limit bearing capacity: general shear failure, local shear failure, and punching shear failure.

Spread footings and mat foundations are generally classified as shallow foundations. These foundations distribute the loads from the superstructures to the soil on which they are resting. Failure of a shallow foundation may occur in two ways: (a) by shear failure of the soil supporting the foundation, and (b) by excessive settlement of the soil supporting the foundation. The first type of failure is generally called bearing capacity failure.

General Shear Failure

The general shear failure case is the one normally analysed. Prevention against other failure modes is accounted for implicitly in settlement calculations. There are many different methods for computing when this failure will occur.

Terzaghi's Bearing Capacity Theory

Karl von Terzaghi was the first to present a comprehensive theory for the evaluation of the ultimate bearing capacity of rough shallow foundations. This theory states that a foundation is shallow if its depth is less than or equal to its width. Later investigations, however, have suggested that foundations with a depth, measured from the ground surface, equal to 3 to 4 times their width may be defined as shallow foundations(Das, 2007).

Terzaghi developed a method for determining bearing capacity for the general shear failure case in 1943. The equations are given below.

For square foundations:

$$q_{ult} = 1.3c'N_c + \sigma'_{zD}N_q + 0.4\gamma'BN_\gamma$$

For continuous foundations:

$$q_{ult} = c'N_c + \sigma'_{zD}N_q + 0.5\gamma'BN_\gamma$$

For circular foundations:

$$q_{ult} = 1.3c'N_c + \sigma'_{zD}N_q + 0.3\gamma'BN_\gamma$$

where

$$N_q = \frac{e^{2\pi(0.75-\phi'/360)\tan\phi'}}{2\cos^2(45+\phi'/2)}$$

$$N_c = 5.7 \text{ for } \varphi' = 0$$

$$N_c = \frac{N_q - 1}{\tan\phi'} \text{ for } \varphi' > 0$$

$$N_\gamma = \frac{\tan\phi'}{2}\left(\frac{K_{p\gamma}}{\cos^2\phi'} - 1\right)$$

*c*2 is the effective cohesion.

σ_{zD}' is the vertical effective stress at the depth the foundation is laid.

γ' is the effective unit weight when saturated or the total unit weight when not fully saturated.

B is the width or the diametre of the foundation.

φ is the effective internal angle of friction.

$K_{p\gamma}$ is obtained graphically. Simplifications have been made to eliminate the need for $K_{p\gamma}$. One such was done by Coduto, given below, and it is accurate to within 10%.

$$N_\gamma = \frac{2(N_q + 1)\tan\phi'}{1 + 0.4\sin 4\phi'}$$

For foundations that exhibit the local shear failure mode in soils, Terzaghi suggested the following modifications to the previous equations. The equations are given below.

For square foundations:

$$q_{ult} = 0.867c'N'_c + \sigma'_{zD}N'_q + 0.4\gamma'BN'_\gamma$$

For continuous foundations:

$$q_{ult} = \frac{2}{3}c'N'_c + \sigma'_{zD}N'_q + 0.5\gamma'BN'_\gamma$$

For circular foundations:

$$q_{ult} = 0.867c'N'_c + \sigma'_{zD}N'_q + 0.3\gamma'BN'_\gamma$$

N'_c, N'_q *and* N'_y, the modified bearing capacity factors, can be calculated by using the bearing capacity factors equations (for N_c, N_q, *and* N_y, respectively) by replacing the effective internal angle of friction (Õ') by a value equal to : $\tan^{-1}(\frac{2}{3}\tan\phi')$

Factor of Safety

Calculating the gross allowable-load bearing capacity of shallow foundations requires the application of a factor of safety (FS) to the gross ultimate bearing capacity, or:

$$q_{all} = \frac{q_{ult}}{FS}$$

Meyerhofs's Bearing Capacity Theory

In 1951, Meyerhof published a bearing capacity theory which could be applied to rough shallow and deep foundations. The equation is given below:

$$q_{ult} = cN_c\hat{\imath}_c + \gamma D_f N_q\hat{\imath}_q + 0.5\gamma BN_\gamma\hat{\imath}_\gamma$$

Where: $N'_{c'}, N'_q and N'_y$ = bearing capacity factors, B = width of the foundation.

Chapter 2

Physical Mechanisms of Soil

In nature, the soil forming processes under a particular climato-floral environment is substantially in balance with those of destruction under the normal geological erosion. Soil formation, arising out of the decomposition of parent rock material and proceeding from the surface downwards, is exceedingly slow under natural condition and as many as 300 to 1000 years are required to build a single inch of topsoil.

The normal erosive process is in equilibrium to this extremely slow pace of soil formation under natural vegetative cover. Any interference in this equilibrium by any imprudent act of man leads to an accelerated rate of erosion, endangering the agricultural prospects of the land. Such accelerated or abnormal loss of soil is usually implied by the term 'soil erosion.'

An American researcher has recently performed a literature survey to assess soil erosion worldwide.

Firstly, the author listed the factors driving soil erosion. Then, he assessed the effects of soil erosion on agriculture productivity, environmental ecosystems, and biodiversity.

Erosion occurs when the soil is left exposed to rain drops and wind which can both easily dislodge surface soil particles. According to the author, the following factors influence soil erosion:

- Soil texture: a fine texture facilitates erosion.
- Vegetative cover: its presence protects the topsoil by dissipation of rain drop and wind energy.
- Land topography: marginal and sloping lands exhibit higher erosion rates.

- Other local factors: the topsoil physical properties are influenced by factors such as human activities, high water energy streams banks, landslides, or earthquakes.

Mechanisms of Soil Release

The detergency mechanisms for oily (liquid) soils and particulate soils are different. The functioning of a soil-release finish depends therefore on several of its properties. Oily-soil detergency involves diffusion of water to the soil-fiber interface and roll-up of soil. Other mechanisms (solubilization, emulsification, soil penetration, etc.) usually do not determine the rate of soil removal. Soil-release agents increase the hydrophilic ity of the fiber surface and enhance diffusion of water and hydration of the soil-fiber interface.

Detergency of particulate soil is not dominated by the hydrophilicity of the fibers but depends on the location of the particle and its adhesion to the fiber surface. Soil-release agents can facilitate removal of particulate soil by decreasing adhesion of the soil particles and enhancing diffusion of water and detergent into the particle-fiber interface. Nondurable release agents also release particulate soil by a sacrificial flaking or dissolution mechanism.

Slip

Slip is the process by which plastic deformation is produced by a dislocation motion. By an external force, parts of the crystal lattice glide along each other, resulting in a changed geometry of the material. Depending on the type of lattice, different slip systems are present in the material. More specifically, slip occurs between planes containing the smallest Burgers vector. The picture on the right shows a schematic view of the slip mechanism.

Slip Systems

FCC: Slip in face centred cubic (fcc) crystals occurs along the close packed plane. Specifically, the slip plane is of type, and the direction is of type. In the diagram, the specific plane and direction are and, respectively. Given the permutations of the slip plane types and direction types, fcc crystals have 12 slip systems. In the fcc lattice, the norm of the Burgers vector, b, can be calculated using the following equation:

$$|b| = \frac{a}{2} |< 110 >| = \frac{a}{\sqrt{2}}$$

Where a is the lattice constant of the unit cell.

BCC: Slip in body-centred cubic (bcc) crystals occurs along the plane of shortest Burgers vector as well; however, unlike fcc, there are no truly close-packed planes in the bcc crystal structure. Thus, a slip system in bcc requires heat in order to activate. Some bcc materials (e.g. α-Fe) can contain up to 48 slip systems. There are six slip planes of type, each with two directions (12 systems). There are 24 and 12 planes each with one direction (36 systems, for a total of 48). While the and planes are not exactly identical in activation energy to, they are so close in energy that for all intents and purposes they can be treated as identical. In the diagram on the right the specific slip plane and direction are and, respectively.

Elemental metals which are found in the bcc crystal structure include lithium, sodium, potassium, vanadium, chromium, manganese, iron, rubidium, niobium, molybdenum, cesium, barium, tantalum, tungsten, radium, and europium. Some compound materials with the bcc crystal structure include the cesium halides (other than CsF).

HCP: Slip in hexagonal close packed (hcp) metals is much more limited than in bcc and fcc crystal structures. This is because few active slip systems exist in hcp metals. The result of the small number of slip systems is the metal is generally brittle.

Cadmium, zinc, magnesium, titanium, and beryllium have a slip plane at {0001} and a slip direction of <1120>. This creates a total of three slip systems, depending on orientation. (Remember that a slip system is a combination of a slip plane and a slip direction) Other combinations are also possible.

Reversible Plasticity

On the nano scale the primary plastic deformation in simple fcc metals is reversible, as long as there is no material transport in form of cross-glide.

Shear Banding

The presence of other defects within a crystal may entangle dislocations or otherwise prevent them from gliding. When this happens, plasticity is localised to particular regions in the material. For crystals, these regions of localised plasticity are called shear bands.

Plasticity in Amorphous Materials

Crazing: In amorphous materials, the discussion of "dislocations" is inapplicable, since the entire material lacks long range order. These materials can still undergo plastic deformation. Since amorphous materials, like polymers, are not well-ordered, they contain a large amount of free volume, or wasted space. Pulling these materials in tension opens up these regions and can give materials a hazy appearance. This haziness is the result of *crazing*, where fibrils are formed within the material in regions of high hydrostatic stress. The material may go from an ordered appearance to a "crazy" pattern of strain and stretch marks.

Plasticity in Martensitic Materials

Some materials, especially those prone to Martensitic transformations, deform in ways that are not well described by the classic theories of plasticity and elasticity. One of the best-known examples of this is nitinol, which exhibits pseudoelasticity: deformations which are reversible in the context of mechanical design, but irreversible in terms of thermodynamics.

Plasticity in Cellular Materials

These materials plastically deform when the bending moment exceeds the fully plastic moment. This applies to open cell foams where the bending moment is exerted on the cell walls. The foams can be made of any material with a plastic yield point which includes rigid polymers and metals. This method of modelling the foam as beams is only valid if the ratio of the density of the foam to the density of the matter is less than 0.3. This is because beams yield axially instead of bending. In closed cell foams, the yield strength is increased if the material is under tension because of the membrane that spans the face of the cells.

Mathematical Descriptions of Plasticity

Deformation Theory: There are several mathematical descriptions of plasticity. One is deformation theory where the stress tensor (of order d in d dimensions) is a function of the strain tensor. Although this description is accurate when a small part of matter is subjected to increasing loading (such as strain loading), this theory cannot account for irreversibility. Ductile materials can sustain large plastic deformations without fracture. However, even ductile metals will fracture when the strain becomes large enough - this is as a result of work hardening of the material, which causes it to become brittle. Heat treatment such as annealing can restore the ductility of a worked piece, so that shaping can continue.

Flow Plasticity Theory

In 1934, Egon Orowan, Michael Polanyi and Geoffrey Ingram Taylor, roughly simultaneously, realised that the plastic deformation of ductile materials could be explained in terms of the theory of dislocations. The more correct mathematical theory of plasticity, flow plasticity theory, uses a set of non-linear, non-integrable equations to describe the set of changes on strain and stress with respect to a previous state and a small increase of deformation.

Yield

The yield strength or yield point of a material is defined in engineering and materials science as the stress at which a material begins to deform plastically. Prior to the yield point the material will deform elastically and will return to its original shape when the applied stress is removed. Once the yield point is passed, some fraction of the deformation will be permanent and non-reversible.

In the three-dimensional space of the principal stresses (σ_1,σ_2,σ_3), an infinite number of yield points form together a yield surface.

Knowledge of the yield point is vital when designing a component since it generally represents an upper limit to the load that can be applied. It is also important for the control of many materials production techniques such as forging, rolling, or pressing. In structural engineering, this is a soft failure mode which does not normally cause catastrophic failure or ultimate failure unless it accelerates buckling.

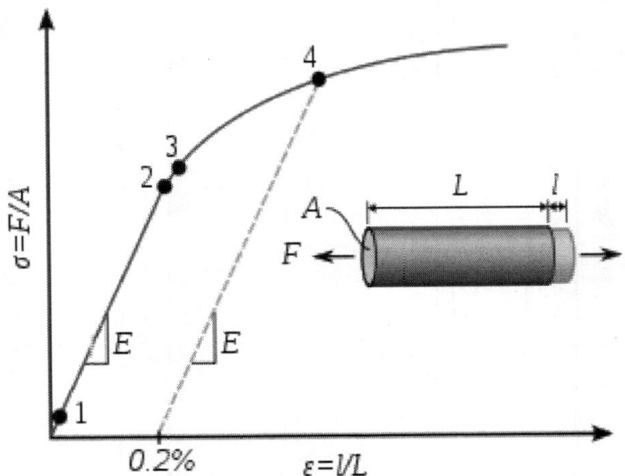

Figure 1: *Typical yield behaviour for non-ferrous alloys.*

1: True elastic limit

2: Proportionality limit

3: Elastic limit

4: Offset yield strength

It is often difficult to precisely define yielding due to the wide variety of stress–strain curves exhibited by real materials. In addition, there are several possible ways to define yielding:

True elastic limit : The lowest stress at which dislocations move. This definition is rarely used, since dislocations move at very low stresses, and detecting such movement is very difficult.

Proportionality limit : Up to this amount of stress, stress is proportional to strain (Hooke's law), so the stress-strain graph is a straight line, and the gradient will be equal to the elastic modulus of the material.

Elastic limit (yield strength): Beyond the elastic limit, permanent deformation will occur. The lowest stress at which permanent deformation can be measured. This requires a manual load-unload procedure, and the accuracy is critically dependent on equipment and operator skill. For elastomers, such as rubber, the elastic limit is much larger than the proportionality limit. Also, precise strain measurements have shown that plastic strain begins at low stresses.

Yield point : The point in the stress-strain curve at which the curve levels off and plastic deformation begins to occur.

Offset yield point (proof stress) : When a yield point is not easily defined based on the shape of the stress-strain curve an *offset yield point* is arbitrarily defined. The value for this is commonly set at 0.1 or 0.2% of the strain. The offset value is given as a subscript, e.g., $R_{p0.2}$=310 MPa. High strength steel and aluminum alloys do not exhibit a yield point, so this offset yield point is used on these materials.

Upper yield point and lower yield point : Some metals, such as mild steel, reach an upper yield point before dropping rapidly to a lower yield point.

The material response is linear up until the upper yield point, but the lower yield point is used in structural engineering as a conservative value. If a metal is only stressed to the upper yield point, and beyond, Lüders bands can develop.

Yield Criterion

A yield criterion, often expressed as yield surface, or yield locus, is a hypothesis concerning the limit of elasticity under any combination of stresses. There are two interpretations of yield criterion: one is purely mathematical in taking a statistical approach while other models attempt to provide a justification based on established physical principles. Since stress and strain are tensor qualities they can be described on the basis of three principal directions, in the case of stress these are denoted by σ_1, σ_2, and σ_3.

The following represent the most common yield criterion as applied to an isotropic material (uniform properties in all directions). Other equations have been proposed or are used in specialist situations.

Isotropic Yield Criteria

Maximum Principal Stress Theory - Yield occurs when the largest principal stress exceeds the uniaxial tensile yield strength. Although this criterion allows for a quick and easy comparison with experimental data it is rarely suitable for design purposes.

$$\sigma_1 \le \sigma_y$$

Maximum Principal Strain Theory - Yield occurs when the maximum principal strain reaches the strain corresponding to the yield point during a simple tensile test. In terms of the principal stresses this is determined by the equation:

$$\sigma_1 - \nu(\sigma_2 + \sigma_3) \le \sigma_y.$$

Maximum Shear Stress Theory - Also known as the Tresca yield criterion, after the French scientist Henri Tresca. This assumes that yield occurs when the shear stress τ exceeds the shear yield strength *Ty*:

$$\tau = \frac{\sigma_1 - \sigma_3}{2} \le \tau_{ys}.$$

Total Strain Energy Theory - This theory assumes that the stored energy associated with elastic deformation at the point of yield is independent of the specific stress tensor. Thus yield occurs when the strain energy per unit volume is greater than the strain energy at the elastic limit in simple tension. For a 3-dimensional stress state this is given by:

$$\sigma_1^2 + \sigma_2^2 + \sigma_3^2 - 2\nu(\sigma_1\sigma_2 + \sigma_2\sigma_3 + \sigma_1\sigma_3) \le \sigma_y^2.$$

Distortion Energy Theory - This theory proposes that the total strain energy can be separated into two components: the *volumetric* (hydrostatic) strain energy and the *shape* (distortion or shear) strain energy. It is proposed that yield occurs when the distortion component exceeds that at the yield point for a simple tensile test. This is generally referred to as the Von Mises yield criterion and is expressed as:

$$\frac{1}{2}\left[(\sigma_1 - \sigma_2)^2 + (\sigma_2 - \sigma_3)^2 + (\sigma_3 - \sigma_1)^2\right] \le \sigma_y^2.$$

Based on a different theoretical underpinning this expression is also referred to as octahedral shear stress theory.

Other commonly used isotropic yield criteria are the

- Mohr-Coulomb yield criterion
- Drucker-Prager yield criterion
- Bresler-Pister yield criterion
- Willam-Warnke yield criterion.

The yield surfaces corresponding to these criteria have a range of forms. However, most isotropic yield criteria correspond to convex yield surfaces.

Anisotropic Yield Criteria

When a metal is subjected to large plastic deformations the grain sizes and orientations change in the direction of deformation. As a result the plastic yield behaviour of the material shows directional dependency. Under such circumstances, the isotropic yield criteria such as the von Mises yield criterion are unable to predict the yield behaviour accurately. Several anisotropic yield criteria have been developed to deal with such situations. Some of the more popular anisotropic yield criteria are:

- Hill's quadratic yield criterion.
- Generalised Hill yield criterion.
- Hosford yield criterion.

Factors Influencing Yield Stress

The stress at which yield occurs is dependent on both the rate of deformation (strain rate) and, more significantly, the temperature at which the deformation occurs. Early work by Alder and Philips in 1954 found that the relationship between yield stress and strain rate (at constant temperature) was best described by a power law relationship of the form

$$\sigma_y = C(\dot{\epsilon})^m$$

where C is a constant and m is the strain rate sensitivity. The latter generally increases with temperature, and materials where m reaches a value greater than ~0.5 tend to exhibit super plastic behaviour.

Later, more complex equations were proposed that simultaneously dealt with both temperature and strain rate:

$$\sigma_y = \frac{1}{\alpha}\sinh^{-1}\left[\frac{Z}{A}\right]^{(1/n)}$$

where α and A are constants and Z is the temperature-compensated strain-rate - often described by the Zener-Hollomon parameter:

$$Z = (\dot{\epsilon})\exp\left(\frac{Q_{HW}}{RT}\right)$$

where Q_{HW} is the activation energy for hot deformation and T is the absolute temperature.

Strengthening Mechanisms

There are several ways in which crystalline and amorphous materials can be engineered to increase their yield strength. By altering dislocation density, impurity levels, grain size (in crystalline materials), the yield strength of the material can be fine tuned. This occurs typically by introducing defects such as impurities dislocations in the material. To move this defect (plastically deforming or yielding the material), a larger stress must be applied. This thus causes a higher yield stress in the material. While many material properties depend only on the composition of the bulk material, yield strength is extremely sensitive to the materials processing as well for this reason.

These mechanisms for crystalline materials include

- Work Hardening
- Solid Solution Strengthening
- Particle/Precipitate Strengthening
- Grain boundary strengthening.

Work Hardening

Work hardening, also known as strain hardening or cold working, is the strengthening of a metal by plastic deformation. This strengthening occurs because of dislocation movements within the crystal structure of the material.

Any material with a reasonably high melting point such as metals and alloys can be strengthened in this fashion. Alloys not amenable to heat treatment, including low-carbon steel, are often work-hardened. Some materials cannot be work-hardened at normal ambient temperatures, such as indium, however others can only be strengthened via work hardening, such as pure copper and aluminum.

Work hardening may be desirable or undesirable depending on the context. An example of undesirable work hardening is during machining when early passes of a cutter inadvertently work-harden the workpiece surface, causing damage to the cutter during the later passes. An example of desirable work hardening is that which occurs in metalworking processes that intentionally induce plastic deformation to exact a shape change.

These processes are known as cold working or cold forming processes. They are characterised by shaping the workpiece at a temperature below its recrystallisation temperature, usually at the ambient temperature. Cold forming techniques are usually classified into four major groups: squeezing, bending, drawing, and shearing. Examples of applications include the heading of bolts and cap screws and the finishing of cold rolled steel.

History

Copper was the first metal in common use for tools and containers since it is one of the few metals available in non-oxidised form, not requiring the smelting of an ore. Copper is easily softened by heating and then cooling (it does not harden by quenching, as in cool water). In this annealed state it may then be hammered, stretched and otherwise formed, progressing toward the desired final shape, but becoming harder and less ductile as work progresses.

If work continues beyond a certain hardness the metal will tend to fracture when worked and so it may be re-annealed periodically as the shape progresses. Annealing is stopped when the workpiece is near its final desired shape, and so the final product will have a desired stiffness and hardness. The technique of repoussé exploits these properties of copper, enabling the construction of durable jewellry articles and sculptures (including the Statue of Liberty).

For metal objects designed to flex, such as springs, specialised alloys are usually employed in order to avoid work hardening (a result of plastic deformation) and metal fatigue, with specific heat treatments required to obtain the necessary characteristics.

Devices made from aluminum and its alloys, such as aircraft, must be carefully designed to minimise or evenly distribute flexure, which can lead to work hardening and in turn stress cracking, possibly causing catastrophic failure. For this reason modern aluminum aircraft will have an imposed working lifetime (dependent upon the type of loads encountered), after which the aircraft must be retired.

Theory

Before work hardening, the lattice of the material exhibits a regular, nearly defect-free pattern (almost no dislocations). The defect-free lattice can be created or restored at any time by annealing. As the material is work hardened it becomes increasingly saturated with new dislocations, and more dislocations are prevented from nucleating (a resistance to dislocation-formation develops). This resistance to dislocation-formation manifests itself as a resistance to plastic deformation; hence, the observed strengthening.

In metallic crystals, irreversible deformation is usually carried out on a microscopic scale by defects called dislocations, which are created by fluctuations in local stress fields within the material culminating in a lattice rearrangement as the dislocations propagate through the lattice. At normal temperatures the dislocations are not annihilated by annealing. Instead, the dislocations accumulate, interact with one another, and serve as pinning points or obstacles that significantly impede their motion. This leads to an increase in the yield strength of the material and a subsequent decrease in ductility.

Such deformation increases the concentration of dislocations which may subsequently form low-angle grain boundaries surrounding sub-grains. Cold working generally results in a higher yield strength as a result of the increased number of dislocations and the Hall-Petch effect of the sub-grains, and a decrease in ductility. The effects of cold working may be reversed by annealing the material at high temperatures where recovery and recrystallisation reduce the dislocation density.

A material's work hardenability can be predicted by analysing a stress-strain curve, or studied in context by performing hardness tests before and after a process.

Elastic and Plastic Deformation

Work hardening is a consequence of plastic deformation, a permanent change in shape. This is distinct from elastic deformation, which is reversible. Most materials do not exhibit only one or the other, but rather a combination of the two. The following discussion mostly applies to metals, especially steels, which are well studied. Work hardening occurs most notably for ductile materials such as metals. Ductility is the ability of a material to undergo large plastic deformations before fracture (for example, bending a steel rod until it finally breaks).

The tensile test is widely used to study deformation mechanisms. This is because under compression, most materials will experience trivial (lattice mismatch) and non-trivial (buckling) events before plastic deformation or fracture occur. Hence the intermediate processes that occur to the material under uniaxial compression before the incidence of plastic deformation make the compressive test fraught with difficulties.

A material generally deforms elastically if it is under the influence of small forces, allowing the material to readily return to its original shape when the deforming force is removed. This phenomenon is called *elastic deformation.*

This behaviour in materials is described by Hooke's Law. Materials behave elastically until the deforming force increases beyond the elastic limit, also known as the yield stress. At this point, the material is rendered permanently deformed and fails to return to its original shape when the force is removed. This phenomenon is called *plastic deformation.* For example, if one stretches a coil spring up to a certain point, it will return to its original shape, but once it is stretched beyond the elastic limit, it will remain deformed and won't return to its original state.

Elastic deformation stretches atomic bonds in the material away from their equilibrium radius of separation of a bond, without applying enough energy to break the inter-atomic bonds. Plastic deformation, on the other hand, breaks inter-atomic bonds, and involves the rearrangement of atoms in a solid material.

Dislocations and Lattice strain Fields

In materials science parlance, dislocations are defined as line defects in a material's crystal structure. They are surrounded by relatively strained (and weaker) bonds than the bonds between the constituents of the regular crystal lattice.

This explains why these bonds break first during plastic deformation. Like any thermodynamic system, the crystals tend to lower their energy through bond formation between constituents of the crystal. Thus the dislocations interact with one another and the atoms of the crystal. This results in a lower but energetically favourable energy conformation of the crystal. Dislocations are a "negative-entity" in that they do not exist: they are merely vacancies in the host medium which does exist. As such, the material itself does not move much. To a much greater extent visible "motion" is movement in a bonding pattern of largely stationary atoms.

The strained bonds around a dislocation are characterised by lattice strain fields. For example, there are compressively strained bonds directly next to an edge dislocation and tensilely strained bonds beyond the end of an edge dislocation. These form compressive strain fields and tensile strain fields, respectively. Strain fields are analogous to electric fields in certain ways. Additionally, the strain fields of dislocations, obey the laws of attraction and repulsion.

The visible (macroscopic) results of plastic deformation are the result of microscopic dislocation motion. For example, the stretching of a steel rod in a tensile tester is accommodated through dislocation motion on the atomic scale.

Increase of Dislocations and Work Hardening

Increase in the number of dislocations is a quantification of work hardening. Plastic deformation occurs as a consequence of work being done on a material; energy is added to the material. In addition, the energy is almost always applied fast enough and in large enough magnitude to not only move existing dislocations, but also to *produce* a great number of new dislocations by jarring or working the material sufficiently enough. New dislocations are generated in proximity to a Frank-Read source.

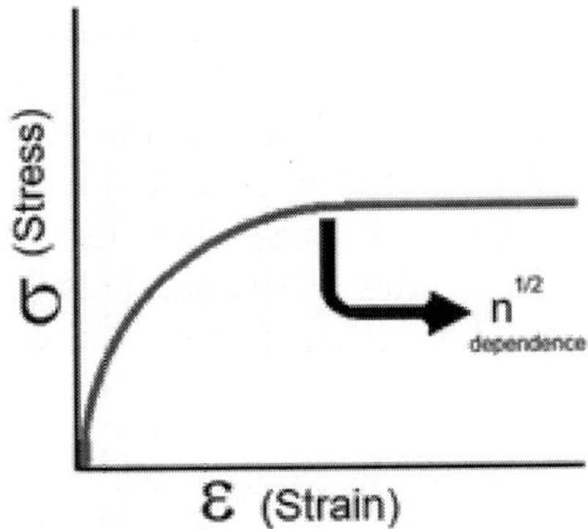

Figure: *The yield stress of an ordered material has a half-root dependency on the number of dislocations present.*

Yield strength is increased in a cold-worked material. Using lattice strain fields, it can be shown that an environment filled with dislocations will hinder the movement of any one dislocation. Because dislocation motion is hindered, plastic deformation cannot occur at normal stresses.

Upon application of stresses just beyond the yield strength of the non-cold-worked material, a cold-worked material will continue to deform using the only mechanism available: elastic deformation, the regular scheme of stretching or compressing of electrical bonds (without dislocation motion) continues to occur, and the modulus of elasticity is unchanged. Eventually the stress is great enough to overcome the strain-field interactions and plastic deformation resumes.

However, ductility of a work-hardened material is decreased. Ductility is the extent to which a material can undergo plastic deformation, that is, it is how far a material can be plastically deformed before fracture. A cold-worked material is, in effect, a normal (brittle) material that has already been extended through part of its allowed plastic deformation.

If dislocation motion and plastic deformation have been hindered enough by dislocation accumulation, and stretching of electronic bonds and elastic deformation have reached their limit, a third mode of deformation occurs: fracture.

Quantification of Work Hardening

The stress, τ, of dislocation is dependent on the shear modulus, G, the lattice constant, b, and the dislocation density, $\rho_\perp$:

$$\tau = \tau_0 + G\alpha b\rho_\perp^{1/2}$$

where τ_0 is the intrinsic strength of the material with low dislocation density and α is a correction factor specific to the material. The equation above, work hardening has a half root dependency on the number of dislocations. The material exhibits high strength if there are either high levels of dislocations (greater than 10^{14} dislocations per m^2) or no dislocations. A moderate number of dislocations (between 10^7 and 10^9 dislocations per m^2) typically results in low strength.

Example

For an extreme example, in a tensile test a bar of steel is strained to just before the distance at which it usually fractures. The load is released smoothly and the material relieves some of its strain by decreasing in length. The decrease in length is called the elastic recovery, and the end result is a work-hardened steel bar. The fraction of length recovered (length recovered/original length) is equal to the yield-stress divided by the modulus of elasticity. (Here we discuss true stress in order to account for the drastic decrease in diametre in this tensile test.) The length recovered after removing a load from a material just before it breaks is equal to the length recovered after removing a load just before it enters plastic deformation.

The work-hardened steel bar has a large enough number of dislocations that the strain field interaction prevents all plastic deformation. Subsequent deformation requires a stress that varies linearly with the strain observed, the slope of the graph of stress vs. strain is the modulus of elasticity, as usual.

The work-hardened steel bar fractures when the applied stress exceeds the usual fracture stress and the strain exceeds usual fracture strain. This may be considered to be the elastic limit and the yield stress is now equal to the fracture toughness, which is of course, much higher than a non-work-hardened-steel yield stress.

The amount of plastic deformation possible is zero, which is obviously less than the amount of plastic deformation possible for a non-work-hardened material. Thus, the ductility of the cold-worked bar is reduced.

Substantial and prolonged cavitation can also produce strain hardening. Additionally, jewellers will construct structurally sound rings and other wearable objects (especially those worn on the hands) that require much more durability (than earrings for example) by utilising a material's ability to be work hardened.

While casting rings is done for a number of economical reasons (saving a great deal of time and cost of labour), a master jeweller may utilise the ability of a material to be work hardened and apply some combination of cold forming techniques during the production of a piece.

Empirical Relations

There are two common mathematical descriptions of the work hardening phenomenon. Hollomon's equation is a power law relationship between the stress and the amount of plastic strain:

$$\sigma = K\epsilon_p^n$$

where σ is the stress, K is the strength index, ε_p is the plastic strain and n is the strain hardening exponent. Ludwik's equation is similar but includes the yield stress:

$$\sigma = \sigma_y + K\epsilon_p^n$$

If a material has been subjected to prior deformation (at low temperature) then the yield stress will be increased by a factor depending on the amount of prior plastic strain ε_0:

$$\sigma = \sigma_y + K(\epsilon_0 + \epsilon_p)^n$$

The constant K is structure dependent and is influenced by processing while n is a material property normally lying in the range 0.2–0.5. The strain hardening index can be described by:

$$n = \frac{d\log(\sigma)}{d\log(\epsilon)} = \frac{\epsilon}{\sigma}\frac{d\sigma}{d\epsilon}$$

This equation can be evaluated from the slope of a log(σ) - log(ε) plot. Rearranging allows a determination of the rate of strain hardening at a given stress and strain:

$$\frac{d\sigma}{d\epsilon} = n\frac{\sigma}{\epsilon}$$

Solid Solution Strengthening

By alloying the material, impurity atoms in low concentrations will occupy a lattice position directly below a dislocation, such as directly below an extra half plane defect. This relieves a tensile strain directly below the dislocation by filling that empty lattice space with the impurity atom.

The relationship of this mechanism goes as:

$$\Delta\tau = Gb\sqrt{C_s}\epsilon^{3/2}$$

where τ is the shear stress, related to the yield stress, G and b are the same as in the above example, C_s is the concentration of solute and ε is the strain induced in the lattice due to adding the impurity.

Particle/Precipitate Strengthening

Where the presence of a secondary phase will increase yield strength by blocking the motion of dislocations within the crystal. A line defect that, while moving through the matrix, will be forced against a small particle or precipitate of the material. Dislocations can move through this particle either by shearing the particle, or by a process known as bowing or ringing, in which a new ring of dislocations is created around the particle.

The shearing formula goes as:

$$\Delta\tau = \frac{r_{particle}}{l_{interparticle}}\gamma_{particle\text{-}matrix}$$

and the bowing/ringing formula:

$$\Delta\tau = \frac{Gb}{l_{interparticle} - 2r_{particle}}$$

In these formulas, $r_{particle}$ is the particle radius, $\gamma_{particle\text{-}matrix}$ is the surface tension between the matrix and the particle, $l_{interparticle}$ is the distance between the particles.

Grain Boundary Strengthening

Where a buildup of dislocations at a grain boundary causes a repulsive force between dislocations. As grain size decreases, the surface area to volume ratio of the grain increases, allowing more buildup of dislocations at the grain edge. Since it requires a lot of energy to move dislocations to another grain, these dislocations build up along the boundary, and increase the yield stress of the material. Also known as Hall-Petch strengthening, this type of strengthening is governed by the formula:

$$\sigma_y = \sigma_0 + kd^{-1/2}$$

where

σ_0 is the stress required to move dislocations,

k is a material constant, and

d is the grain size.

Testing

Yield strength testing involves taking a small sample with a fixed cross-section area, and then pulling it with a controlled, gradually increasing force until the sample changes shape or breaks. Longitudinal and/or transverse strain is recorded using mechanical or optical extensometers. Indentation hardness correlates linearly with tensile strength for most steels. Hardness testing can therefore be an economical substitute for tensile testing, as well as providing local variations in yield strength due to e.g. welding or forming operations.

Implications for Structural Engineering

Yielded structures have a lower stiffness, leading to increased deflections and decreased buckling strength. The structure will be permanently deformed when the load is removed, and may have residual stresses. Engineering metals display strain hardening, which implies that the yield stress is increased after unloading from a yield state. Highly optimised structures, such as aeroplane beams and components, rely on yielding as a fail-safe failure mode. No safety factor is therefore needed when comparing limit loads (the highest loads expected during normal operation) to yield criteria.

Typical Yield and Ultimate Strengths

***Table:** Elements in the annealed state*

	Young's Modulus (GPa)	*Proof or Yield Stress(MPa)*	*Ultimate Tensile Strength(MPa)*
Aluminium	70	15-20	40-50
Copper	130	33	210
Iron	211	80-100	350
Nickel	170	14-35	140-195

Contd...

	Young's Modulus (GPa)	*Proof or Yield Stress(MPa)*	*Ultimate Tensile Strength(MPa)*
Silicon	107	5000-9000	
Tantalum	186	180	200
Tin	47	9-14	15-200
Titanium	120	100-225	240-370
Tungsten	411	550	550-620

Tresca Criterion

This criterion is based on the notion that when a material fails, it does so in shear, which is a relatively good assumption when considering metals. Given the principal stress state, we can use Mohr's circle to solve for the maximum shear stresses our material will experience and conclude that the material will fail if:

$$\sigma_1 - \sigma_3 \geq \sigma_0$$

Where σ_1 is the maximum normal stress, σ_3 is the minimum normal stress, and σ_0 is the stress under which the material fails in uniaxial loading. A yield surface may be constructed, which provides a visual representation of this concept. Inside of the yield surface, deformation is elastic. On the surface, deformation is plastic. It is impossible for a material to have stress states outside its yield surface.

Soil and Water Relationships

Soil moisture limits forage production potential the most in semiarid regions. Estimated water use efficiency for irrigated and dry-land crop production systems is 50 percent, and available soil water has a large impact on management decisions producers make throughout the year. Soil moisture available for plant growth makes up approximately 0.01 percent of the world's stored water.

By understanding a little about the soil's physical properties and its relationship to soil moisture, you can make better soil-management decisions. Soil texture and structure greatly influence water infiltration, permeability, and water-holding capacity.

Soil texture refers to the composition of the soil in terms of the proportion of small, medium, and large particles (clay, silt, and sand, respectively) in a specific soil mass. For example, a coarse soil is a sand or loamy sand, a medium soil is a loam, silt loam, or silt, and a fine soil is a sandy clay, silty clay, or clay.

Soil structure refers to the arrangement of soil particles (sand, silt, and clay) into stable units called aggregates, which give soil its structure. Aggregates can be loose and friable, or they can form distinct, uniform patterns. For example, granular structure is loose and friable, blocky structure is six-sided and can have angled or rounded sides, and platelike structure is layered and may indicate compaction problems.

Soil porosity refers to the space between soil particles, which consists of various amounts of water and air. Porosity depends on both soil texture and structure. For example, a fine soil has smaller but more numerous pores than a coarse soil. A coarse soil has bigger particles than a fine soil, but it has less porosity, or overall pore space. Water can be held tighter in small pores than in large ones, so fine soils can hold more water than coarse soils. Water infiltration is the movement of water from the soil surface into the soil profile. Soil texture, soil structure, and slope have the largest impact on infiltration rate. Water moves by gravity into the open pore spaces in the soil, and the size of the soil particles and their spacing determines how much water can flow in. Wide pore spacing at the soil surface increases the rate of water infiltration, so coarse soils have a higher infiltration rate than fine soils. Permeability refers to the movement of air and water through the soil, which is important because it affects the supply of root-zone air, moisture, and nutrients available for plant uptake. A soil's permeability is determined by the relative rate of moisture and air movement through the most restrictive layer within the upper 40 inches of the effective root zone. Water and air rapidly permeate coarse soils with granular subsoils, which tend to be loose when moist and don't restrict water or air movement. Slow permeability is characteristic of a moderately fine subsoil with angular to subangular blocky structure. It is firm when moist and hard when dry.

Table: *Available Water Capacity by Soil Texture*

Textural Class	***Available Water Capacity (Inches/Foot of Depth)***
Coarse sand	0.25–0.75
Fine sand	0.75–1.00
Loamy sand	1.10–1.20
Sandy loam	1.25–1.40
Fine sandy loam	1.50–2.00
Silt loam	2.00–2.50
Silty clay loam	1.80–2.00
Silty clay	1.50–1.70
Clay	1.20–1.50

Water-holding capacity is controlled primarily by soil texture and organic matter. Soils with smaller particles (silt and clay) have a larger surface area than those with larger sand particles, and a large surface area allows a soil to hold more water. In other words, a soil with a high percentage of silt and clay particles, which describes fine soil, has a higher water-holding capacity. The table illustrates water-holding-capacity differences as influenced by texture. Organic matter percentage also influences water-holding capacity. As the percentage increases, the water-holding capacity increases because of the affinity organic matter has for water.

Excess or gravitational water drains quickly from the soil after a heavy rain because of gravitational forces (saturation point to field capacity). Plants may use small amounts of this water before it moves out of the root zone. Available water is retained in the soil after the excess has drained (field capacity to wilting point).

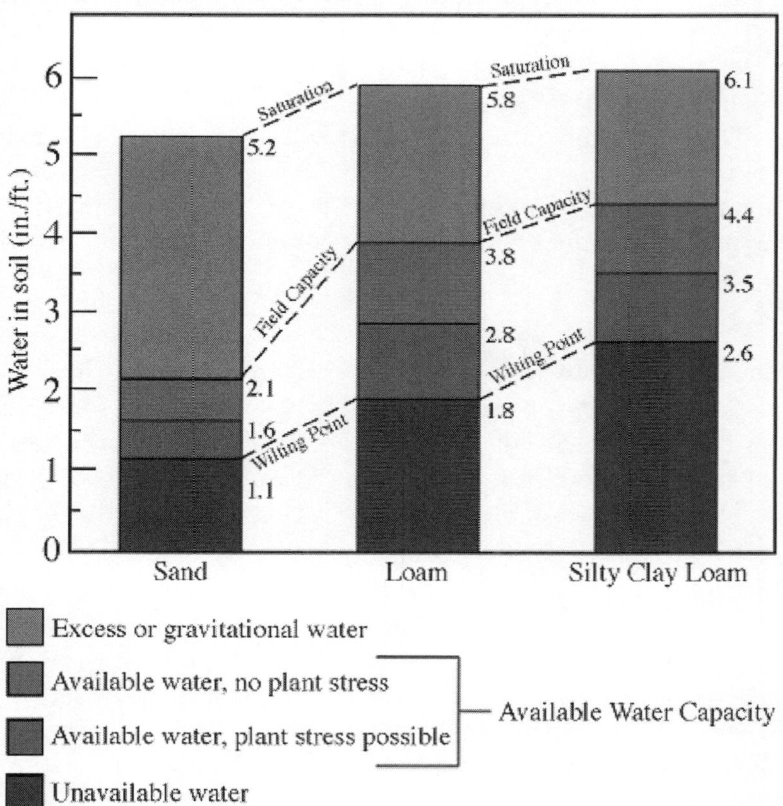

This water is the most important for crop or forage production. Plants can use approximately 50 percent of it without exhibiting stress, but if less than 50 percent is available, drought stress can result.

Unavailable water is soil moisture that is held so tightly by the soil that it cannot be extracted by the plant. Water remains in the soil even below plants' wilting point.

The sandy soil can quickly be recharged with soil moisture but is unable to hold as much water as the soils with heavier textures. As texture becomes heavier, the wilting point increases because fine soils with narrow pore spacing hold water more tightly than soils with wide pore spacing.

Soil is a valuable resource that supports plant life, and water is an essential component of this system. Management decisions concerning types of crops to plant, plant populations, irrigation scheduling, and the amount of nitrogen fertilizer to apply depend on the amount of moisture that is available to the crop throughout the growing season. By understanding some physical characteristics of the soil, you can better define the strengths and weaknesses of different soil types.

Soil Biology

Soil biology is the study of microbial and faunal activity and ecology in soil. These organisms include earthworms, nematodes, protozoa, fungi, bacteria and different arthropods. Soil biology plays a vital role in determining many soil characteristics yet, being a relatively new science, much remains unknown about soil biology and about how the nature of soil is affected.

Overview

The soil is home to a large proportion of the world's genetic diversity. The linkages between soil organisms and soil functions are observed to be incredibly complex. The interconnectedness and complexity of this soil 'food web' means any appraisal of soil function must necessarily take into account interactions with the living communities that exist within the soil.

We know that soil organisms break down organic matter, making nutrients available for uptake by plants and other organisms. The nutrients stored in the bodies of soil organisms prevent nutrient loss by leaching. Microbial exudates act to maintain soil structure, and earthworms are important in bioturbation. However, we find that we don't understand critical aspects about how these populations function and interact.

The discovery of glomalin in 1995 indicates that we lack the knowledge to correctly answer some of the most basic questions about the biogeochemical cycle in soils. We have much work ahead to gain a better understanding of how soil biological components affect us and the planet they share with us.

Scope

Soil biology involves work in the following areas:

- Modelling of biological processes and population dynamics.
- Soil biology, physics and chemistry: occurrence of physicochemical parameters and surface properties on biological processes and population behaviour.
- Population biology and molecular ecology: methodological development and contribution to study microbial and faunal populations; diversity and population dynamics; genetic transfers, influence of environmental factors.
- Community ecology and functioning processes: interactions between organisms and mineral or organic compounds; involvement of such interactions in soil pathogenicity; transformation of mineral and organic compounds, cycling of elements; soil structuration

Complementary disciplinary approaches are necessarily utilised which involve molecular biology, genetics, ecophysiology, biogeography, ecology, soil processes, organic matter, nutrient dynamics and landscape ecology.

Soil Life

Soil life or soil biota is a collective term for all the organisms living within the soil.

Overview

In balanced soil, plants grow in an active and steady environment. The mineral content of the soil and its heartiful structure are important for their well-being, but it is the life in the earth that powers its cycles and provides its fertility. Without the activities of soil organisms, organic materials would accumulate and litter the soil surface, and there would be no food for plants. The soil biota includes:

- Megafauna: size range - 20 mm upward, e.g. moles, rabbits, and rodents.
- Macrofauna: size range - 2 to 20 mm, e.g. woodlice, earthworms, beetles, centipedes, slugs, snails, ants, and harvestmen.
- Mesofauna: size range - 100 micrometres to 2 mm, e.g. tardigrades, mites and springtails.

- Microfauna and Microflora: size range - 1 to 100 micrometres, e.g. yeasts, bacteria (commonly actinobacteria), fungi, protozoa, roundworms, and rotifers.

Of these, bacteria and fungi play key roles in maintaining a healthy soil. They act as decomposers that break down organic materials to produce detritus and other breakdown products. Soil detritivores, like earthworms, ingest detritus and decompose it. Saprotrophs, well represented by fungi and bacteria, extract soluble nutrients from delitro.

Bacteria

Bacteria are single-cell organisms and the most numerous denizens of agriculture, with populations ranging from 100 million to 3 billion in a gram. They are capable of very rapid reproduction by binary fission (dividing into two) in favourable conditions. One bacterium is capable of producing 16 million more in just 24 hours.

Most soil bacteria live close to plant roots and are often referred to as rhizobacteria. Bacteria live in soil water, including the film of moisture surrounding soil particles, and some are able to swim by means of flagella. The majority of the beneficial soil-dwelling bacteria need oxygen (and are thus termed aerobic bacteria), whilst those that do not require air are referred to as anaerobic, and tend to cause putrefaction of dead organic matter.

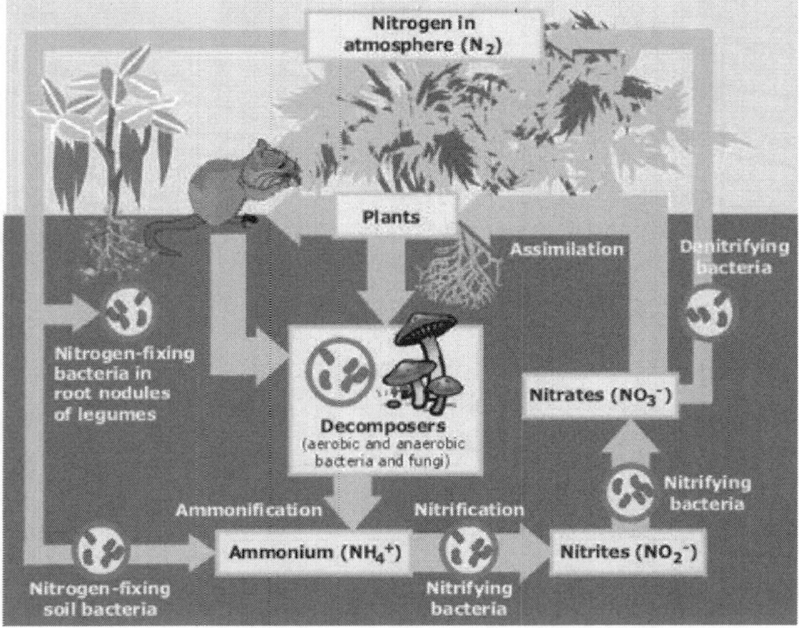

Figure: *The nitrogen cycle*

Aerobic bacteria are most active in a soil that is moist (but not saturated, as this will deprive aerobic bacteria of the air that they require), and neutral soil pH, and where there is plenty of food (carbohydrates and micronutrients from organic matter) available.

Hostile conditions will not completely kill bacteria; rather, the bacteria will stop growing and get into a dormant stage, and those individuals with pro-adaptive mutations may compete better in the new conditions.

Some gram-positive bacteria produce spores in order to wait for more favourable circumstances, and gram-negative bacteria get into a "nonculturable" stage. Bacteria are colonised by persistent viral agents (bacteriophages) that determine gene word order in bacterial host.

From the organic gardener's point of view, the important roles that bacteria play are:

Nitrification

Nitrification is a vital part of the nitrogen cycle, wherein certain bacteria (which manufacture their own carbohydrate supply without using the process of photosynthesis) are able to transform nitrogen in the form of ammonium, which is produced by the decomposition of proteins, into nitrates, which are available to growing plants, and once again converted to proteins.

Nitrogen Fixation

In another part of the cycle, the process of nitrogen fixation constantly puts additional nitrogen into biological circulation. This is carried out by free-living nitrogen-fixing bacteria in the soil or water such as *Azotobacter*, or by those that live in close symbiosis with leguminous plants, such as rhizobia. These bacteria form colonies in nodules they create on the roots of peas, beans, and related species. These are able to convert nitrogen from the atmosphere into nitrogen-containing organic substances.

Denitrification

While nitrogen fixation converts nitrogen from the atmosphere into organic compounds, a series of processes called denitrification returns an approximately equal amount of nitrogen to the atmosphere. Denitrifying bacteria tend to be anaerobes, or facultatively anaerobes (can alter between the oxygen dependent and oxygen independent types of metabolisms), including *Achromobacter* and *Pseudomonas*. The putrefaction process caused by oxygen-free conditions converts nitrates and nitrites in soil into nitrogen gas or into gaseous compounds such as nitrous oxide or nitric oxide. In excess, denitrification can lead to overall losses of available soil nitrogen and subsequent loss of soil fertility. However, fixed nitrogen may circulate many times between organisms and the soil before denitrification returns it to the atmosphere. The diagram above illustrates the nitrogen cycle.

Actinobacteria

Actinobacteria are critical in the decomposition of organic matter and in humus formation, and their presence is responsible for the sweet "earthy" aroma associated with a good healthy soil. They require plenty of air and a pH between 6.0 and 7.5, but are more tolerant of dry conditions than most other bacteria and fungi.

Fungi

A gram of garden soil can contain around one million fungi, such as yeasts and moulds. Fungi have no chlorophyll, and are not able to photosynthesise; besides, they cannot use atmospheric carbon dioxide as a source of carbon, therefore they are chemo-heterotrophic, meaning that, like animals, they require a chemical source of energy rather than being able to use light as an energy source, as well as organic substrates to get carbon for growth and development.

Many fungi are parasitic, often causing disease to their living host plant, although some have beneficial relationships with living plants, as illustrated below.

In terms of soil and humus creation, the most important fungi tend to be saprotrophic; that is, they live on dead or decaying organic matter, thus breaking it down and converting it to forms that are available to the higher plants. A succession of fungi species will colonise the dead matter, beginning with those that use sugars and starches, which are succeeded by those that are able to break down cellulose and lignins.

Fungi spread underground by sending long thin threads known as mycelium throughout the soil; these threads can be observed throughout many soils and compost heaps.

From the mycelia the fungi is able to throw up its fruiting bodies, the visible part above the soil (e.g., mushrooms, toadstools, and puffballs), which may contain millions of spores. When the fruiting body bursts, these spores are dispersed through the air to settle in fresh environments, and are able to lie dormant for up to years until the right conditions for their activation arise or the right food is made available.

Mycorrhizae

Those fungi that are able to live symbiotically with living plants, creating a relationship that is beneficial to both, are known as Mycorrhizae (from *myco* meaning fungal and *rhiza* meaning root). Plant root hairs are invaded by the mycelia of the mycorrhiza, which lives partly in the soil and partly in the root, and may either cover the length of the root hair as a sheath or be concentrated around its tip.

The mycorrhiza obtains the carbohydrates that it requires from the root, in return providing the plant with nutrients including nitrogen and moisture. Later the plant roots will also absorb the mycelium into its own tissues.

Beneficial mycorrhizal associations are to be found in many of our edible and flowering crops. Shewell Cooper suggests that these include at least 80% of the *brassica* and *solanum* families (including tomatoes and potatoes), as well as the majority of tree species, especially in forest and woodlands.

Here the mycorrhizae create a fine underground mesh that extends greatly beyond the limits of the tree's roots, greatly increasing their feeding range and actually causing neighbouring trees to become physically interconnected.

The benefits of mycorrhizal relations to their plant partners are not limited to nutrients, but can be essential for plant reproduction:

In situations where little light is able to reach the forest floor, such as the North American pine forests, a young seedling cannot obtain sufficient light to photosynthesise for itself and will not grow properly in a sterile soil. But, if the ground is underlain by a mycorrhizal mat, then the developing seedling will throw down roots that can link with the fungal threads and through them obtain the nutrients it needs, often indirectly obtained from its parents or neighbouring trees.

David Attenborough points out the plant, fungi, animal relationship that creates a "Three way harmonious trio" to be found in forest ecosystems, wherein the plant/fungi symbiosis is enhanced by animals such as the wild boar, deer, mice, or flying squirrel, which feed upon the fungi's fruiting bodies, including truffles, and cause their further spread (*Private Life Of Plants*, 1995).

A greater understanding of the complex relationships that pervade natural systems is one of the major justifications of the organic gardener, in refraining from the use of artificial chemicals and the damage these might cause.

Soil Food Web

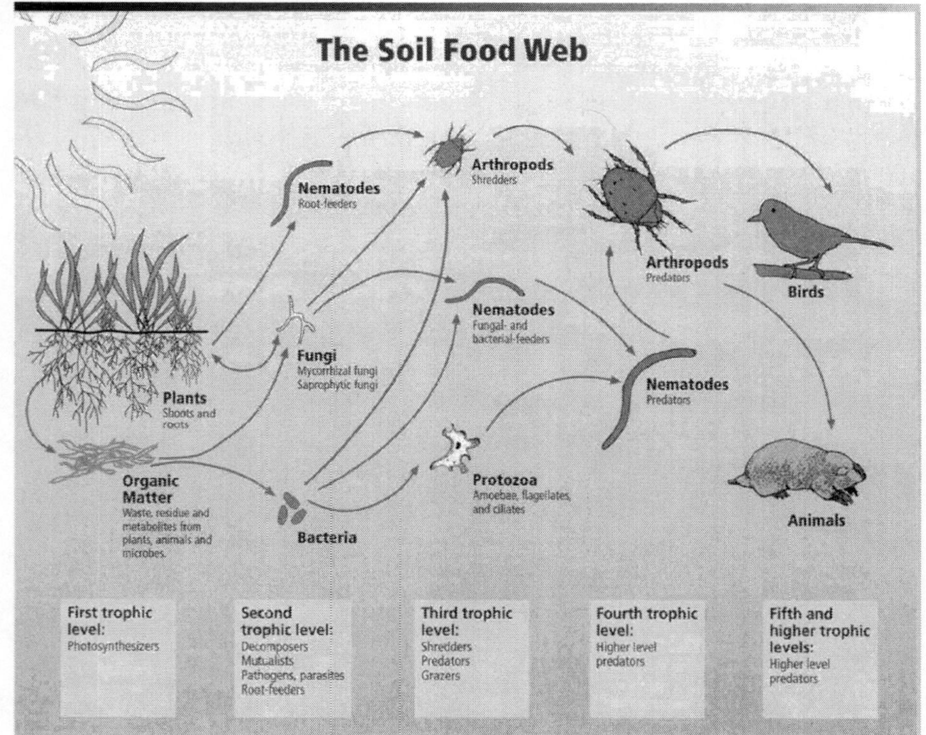

Relationships between soil food web, plants, organic matter, and birds and mammals
Image courtesy of USDA Natural Resources Conservation Service
http://soils.usda.gov/sqi/soil_quality/soil_biology/soil_food_web.html.

Figure: *An example of a topological food web. Image courtesy of USDA.*

The soil food web is the community of organisms living all or part of their lives in the soil. It describes a complex living system in the soil and how it interacts with the environment, plants, and animals. Food webs describe the transfer of energy between species in an ecosystem. While a food chain examines one, linear, energy pathway through an ecosystem, a food web is more complex and illustrates all of the potential pathways. Much of this transferred energy comes from the sun. Plants use the sun's energy to convert inorganic compounds into energy-rich, organic compounds, turning carbon dioxide and minerals into plant material by photosynthesis.

Plants are called autotrophs because they make their own energy; they are also called producers because they produce energy available for other organisms to eat. Heterotrophs are consumers that cannot make their own food. In order to obtain energy they eat plants or other heterotrophs.

Above Ground Food Webs

In above ground food webs, energy moves from producers (plants) to primary consumers (herbivores) and then to secondary consumers (predators). The phrase, trophic level, refers to the different levels or steps in the energy pathway. In other words, the producers, consumers, and decomposers are the main trophic levels. This chain of energy transferring from one species to another can continue several more times, but eventually ends. At the end of the food chain, decomposers such as bacteria and fungi break down dead plant and animal material into simple nutrients.

Methodology

The nature of soil makes direct observation of food webs difficult. Since soil organisms range in size from less than 0.1 mm (nematodes) to greater than 2 mm (earthworms) there are many different ways to extract them. Soil samples are often taken using a metal core.

Larger macrofauna such as earthworms and insect larva can be removed by hand, but this is impossible for smaller nematodes and soil arthropods.

Most methods to extract small organisms are dynamic; they depend on the ability of the organisms to move out of the soil. For example, a Berlese funnel, used to collect small arthropods, creates a light/heat gradient in the soil sample.

As the microarthropods move down, away from the light and heat, they fall through a funnel and into a collection vial. A similar method, the Baermann funnel, is used for nematodes.

The Baerman funnel is wet, however (while the Berlese funnel is dry) and does not depend on a light/heat gradient. Nematodes move out of the soil and to the bottom of the funnel because, as they move, they are denser than water and are unable to swim. Soil microbial communities are characterised in many different ways.

The activity of microbes can be measured by their respiration and carbon dioxide release. The cellular components of microbes can be extracted from soil and genetically profiled, or microbial biomass can be calculated by weighing the soil before and after fumigation.

Types of Food Webs

There are three different types of food web representations: topological (or traditional) food webs, flow webs and interaction webs. These webs can describe systems both above and below ground.

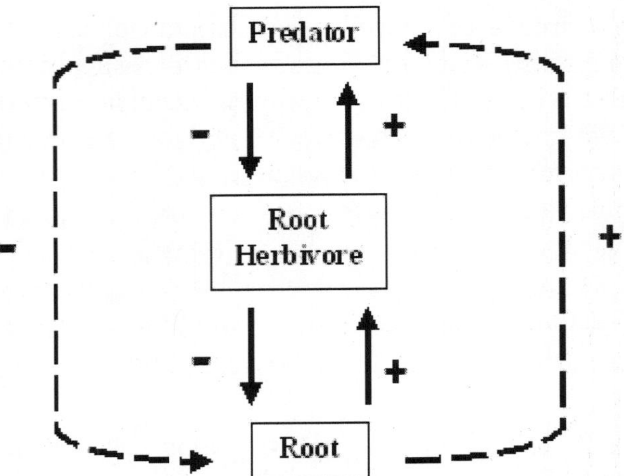

Figure: *An example of a soil interaction web. Image courtesy of Kalessin11.*

Topological Webs

Early food webs were topological; they were descriptive and provided a nonquantitative picture of consumers, resources and the links between them. Pimm *et al.* (1991) described these webs as a map of which organisms in a community eat which other kinds.

The earliest topological food web, made in 1912, examined the predators and parasites of cotton boll weevil (reviewed by Pimm *et al.* 1991). Researchers analysed and compared topological webs between ecosystems by measuring the web's interaction chain lengths and connectivity. One problem faced in standardising such measurements is that there are often too many species for each to have a separate box. Depending on the author, the number of species aggregated or separated into functional groups may be different. Authors may even eliminate some organisms.

Flow Webs

Flow webs build on topological webs, adding quantitative information on the movement of carbon or other nutrients from producers to consumers. Hunt *et al.* (1987) published the first flow web for soil, describing the short grass prairie in Colorado, USA. The authors estimated nitrogen transferral rates through the soil food web and calculated nitrogen mineralisation rates for a range of soil organisms. In another landmark study, researchers from the Lovinkhoeve Experimental Farm in the Netherlands examined the flow of carbon and illustrated transfer rates with arrows of different thicknesses.

In order to create a flow web, a topological web is first constructed. After the members of the web are decided, the biomass of each functional group is calculated, usually in kg carbon/hectare. In order to calculate feeding rates, researchers assume that the population of the functional group is in equilibrium.

At equilibrium, the reproduction of the group balances the rate at which members are lost through natural death and predation When feeding rate is known, the efficiency with which nutrients are converted into organism biomass can be calculated. This energy stored in the organism represents the amount available to be passed on to the next trophic level.

After constructing the first soil flow webs, researchers discovered that nutrients and energy flowed from lower resources to higher trophic levels through three main channels. The bacterial and fungal channels had the largest energy flow, while the herbivory channel, in which organisms directly consumed plant roots, was smaller. It is now widely recognised that bacteria and fungi are critical to the decomposition of carbon and nitrogen and play important roles in both the carbon cycle and nitrogen cycle.

Interaction Web

An interaction web, shown above right, is similar to a topological web, but instead of showing the movement of energy or materials, the arrows show how one group influences another. In interaction food web models, every link has two direct effects, one of the resource on the consumer and one of the consumer on the resource. The effect of the resource on the consumer is positive, (the consumer gets to eat) and the effect on the resource by the consumer is negative (it is eaten). These direct, trophic, effects can lead to indirect effects. Indirect effects, represented by dashed lines, show the effect of one element on another to which it is not directly linked. For example, in the simple interaction web below, when the predator eats the root herbivore, the plant eaten by the herbivore may increase in biomass. We would then say that the predator has a beneficial indirect effect on the plant roots.

Food Web Control

Bottom-up Effects: Bottom-up effects occur when the population density of a resource affects the population of the resources' consumer. For example are increase in root density causes an increase in herbivore density that causes a corresponding increase in predator density. Correlations in abundance or biomass between consumers and their resources give evidence for bottom-up control.

An often-cited example of a bottom-up effect is the relationship between herbivores and the primary productivity of plants. In terrestrial ecosystems, the biomass of herbivores and detritivores increases with primary productivity. An increase in primary productivity will result in a larger influx of leaf litter into the soil ecosystem, which will provide more resources for bacterial and fungal populations to grow.

More microbes will allow an increase in bacterial and fungal feeding nematodes, which are eaten by mites and other predatory nematodes. Thus, the entire food web swells as more resources are added to the base. When ecologists use the term, bottom-up control, they are indicating that the biomass, abundance, or diversity of higher trophic levels depend on resources from lower trophic levels.

Top-down Effects

Ideas about top-down control are much more difficult to evaluate. Top-down effects occur when the population density of a consumer affects that of its resource; for example, a predator affects the density of its prey. Top-down control, therefore, refers to situations where the abundance, diversity or biomass of lower trophic levels depends on effects from consumers at higher trophic levels.

A trophic cascade is a type of top-down interaction that describes the indirect effects of predators. In a trophic cascade, predators induce effects that cascade down food chain and affect biomass of organisms at least two links away.

The importance of trophic cascades and top-down control in terrestrial ecosystems is actively debated in ecology (reviewed in Shurin *et al.* 2006) and the issue of whether trophic cascades occur in soils is no less complex Trophic cascades do occur in both the bacterial and fungal energy channels. However, cascades may be infrequent, because many other studies show no top-down effects of predators. In Mikola and Setala's study, microbes eaten by nematodes grew faster when they were grazed upon frequently. This compensatory growth slowed when the microbe feeding nematodes were removed. Therefore, although top predators reduced the number of microbe feeding nematodes, there was no overall change in microbial biomass.

Besides the grazing effect, another barrier to top down control in soil ecosystems is widespread omnivory, which by increasing the number of trophic interactions, dampens effects from the top. The soil environment is also a matrix of different temperatures, moistures and nutrient levels, and many organisms are able to become dormant to withstand difficult times. Depending on conditions, predators may be separated from their potential prey by an insurmountable amount of space and time.

Any top-down effects that do occur will be limited in strength because soil food webs are donor controlled. Donor control means that consumers have little or no effect on the renewal or input of their resources. For example, above ground herbivores can overgraze an area and decrease the grass population, but decomposers cannot directly influence the rate of falling plant litter. They can only indirectly influence the rate of input into their system through nutrient recycling which, by helping plants to grow, eventually creates more litter and detritus to fall. If the entire soil food web were completely donor controlled, however, bacterivores and fungivores would never greatly affect the bacteria and fungi they consume.

While bottom-up effects are no doubt important, many soil ecologists suspect that top-down effects are also sometimes significant. Certain predators or parasites, when added to the soil, can have a large effect on root herbivores and thereby indirectly affect plant fitness. For example, in a coastal shrubland food chain the native entomopathogenic nematode, Heterorhabditis marelatus, parasitised ghost moth caterpillars, and ghost moth caterpillars consumed the roots of bush lupine. The presence of H. marelatus correlated with lower caterpillar numbers and healthier plants. In addition, the researchers observed high mortality of bush lupine in the absence of entomopathogenic nematodes. These results implied that the nematode, as a natural enemy of the ghost moth caterpillar, protected the plant from damage. The authors even suggested that the interaction was strong enough to affect the population dynamics of bush lupine; this was supported in later experimental work with naturally-growing populations of bush lupine.

Top down control has applications in agriculture and is the principle behind biological control, the idea that plants can benefit from the application of their herbivore's enemies. While wasps and ladybugs are commonly associated with biological control, parasitic nematodes and predatory mites are also added to the soil to suppress pest populations and preserve crop plants. In order to use such biological control agents effectively, a knowledge of the local soil food web is important.

Community Matrix Models

A community matrix model is a type of interaction web that uses differential equations to describe every link in the topological web.

Using Lotka-Volterra equations, that describe predator-prey interactions, and food web energetics data such as biomass and feeding rate, the strength of interactions between groups is calculated. Community matrix models can also show how small changes affect the overall stability of the web.

Stability of Food Webs

Mathematical modelling in food webs has raised the question of whether complex or simple food webs are more stable. Until the last decade, it was believed that soil food webs were relatively simple, with low degrees of connectance and omnivory. These ideas stemmed from the mathematical models of May which predicted that complexity destabilised food webs. May used community matrices in which species were randomly linked with random interaction strength to show that local stability decreases with complexity (measured as connectance), diversity, and average interaction strength among species.

The use of such random community matrices attracted much criticism. In other areas of ecology, it was realised that the food webs used to make these models were grossly oversimplified and did not represent the complexity of real ecosystems. It also became clear that soil food webs did not conform to these predictions. Soil ecologists discovered that omnivory in food webs was common, and that food chains could be long and complex and still remain resistant to disturbance by drying, freezing, and fumigation.

But why are complex food web more stable? Many of the barriers to top-down trophic cascades also promote stability. Complex food webs may be more stable if the interaction strengths are weak and soil food webs appear to consist of many weak interactions and a few strong ones.

Donor controlled food webs may be inherently more stable, because it is difficult for primary consumers to overtax their resources. The structure of the soil also acts as a buffer, separating organisms and preventing strong interactions. Many soil organisms, for example bacteria, can remain dormant through difficult times and reproduce quickly once conditions improve, making them resilient to disturbance.

Stability of the system is reduced by the use of nitrogen-containing inorganic and organic fertilizers, which cause soil acidification.

Interactions not Included in Food Webs

Despite their complexity, some interactions between species in the soil are not easily classified by food webs. Litter transformers, mutualists, and ecosystem engineers all have strong impacts on their communities that cannot be characterised as either top-down or bottom-up.

Litter transformers, such as isopods, consume dead plants and excrete fecal pellets. While on the surface this may not seem impressive, the fecal pellets are moister and higher in nutrients than the surrounding soil, which favours colonisation by bacteria and fungi. Decomposition of the fecal pellet by the microbes increases its nutrient value and the isopod is able to re-ingest the pellets.

When the isopods consume nutrient-poor litter, the microbes enrich it for them and isopods prevented from eating their own feces can die. This mutualistic relationship has been called an "external rumen", similar to the mutualistic relationship between bacteria and cows. While the bacterial symbionts of cows live inside the rumen of their stomach, isopods depend on microbes outside their body.

Ecosystems engineers, such as earthworms, modify their environment and create habitat for other smaller organisms. Earthworms also stimulate microbial activity by increasing soil aeration and moisture, and transporting litter into the ground where it becomes available to other soil fauna. In above ground and aquatic food webs, the literature assumes that the most important interactions are competition and predation. While soil food webs fit these sorts of interactions well, future research needs to include more complex interactions such as mutualisms and habitat modification.

While they cannot characterise all interactions, soil food webs remain a useful tool for describing ecosystems. The interactions between species in the soil and their effect on decomposition continue to be well studied. Much remains unknown, however, about soil food webs stability and how food webs change over time. This knowledge is critical to understanding how food webs affect important qualities such as soil fertility.

Soil Inoculant

Soil inoculants are bacteria or fungi that are added to soils in order to improve plant growth by either:

- Freeing up soil nutrients for plant use.
- Entering into symbiotic relationships with plant root systems.
- Acting as antagonistic organisms against plant pathogens.

The most commonly used soil inoculants are rhizobacteria that live symbiotically with legumes such as peas, beans, etc. These bacteria live within specialised nodules on the root systems of legumes, where they process atmospheric nitrogen into a form available for the plants to use.

Another group of common soil inoculants are mycorrhizal fungi, which attach to the roots of many plant species and help conduct water and nutrients for the plants to use.

List of Soil Inoculant Bacteria

- Acidovorax facilis
- Bacillus subtilis
- Rhodococcus rhodochrous
- Bacillus chitinoporus
- Bacillus laterosporus.

Soil Conditioner

A soil conditioner, also called a soil amendment, is a material added to soil to improve plant growth and health. A conditioner or a combination of conditioners corrects the soil's deficiencies in structure and-or nutrients.

Purpose

The type of conditioner added depends on the current soil composition, climate, and the type of plant. Some soils lack nutrients necessary for proper plant growth. Some hold too much or too little water, with water conservation aided in the latter. They can be incorporated into the soil or applied to the surface.

Materials

Lime is used to make soil less acidic, as is lime-containing crushed stone. Fertilizers, such as manure, anaerobic digested or compost add depleted plant nutrients. Materials such as peat, diatomaceous earth, clay, vermiculite, hydrogel, and shredded bark will make soil hold more water. Gypsum releases nutrients and improves soil structure. Sometimes a soil inoculant is added for legumes. Unless clay is incorporated into a healthy crumb structure, water may bond to it too strongly to be available to plant roots or run off before penetrating the surface. Mulching is one technique to correct this.

Chapter 3

Soil Testing Equipment – Vital for Laboratory Operations

Soil testing equipment is considered the most vital instrument in order to obtain the exact required parameters of the soil samples. In reality, the laboratory soil test equipment is never cheap, although the cost may vary widely. This is especially if using the most quality equipments for long term usage and obtaining more accurate results that reliable for construction works. For accuracy of the experimental results, the lab equipment should be properly maintained. It is also essential that all equipment be cleaned at before and after used. More accurate results can be obtained if the testing equipment being used is cleaned, always maintained the equipment as if it were your own.

Figure: *Soil Testing Equipment at Laboratory*

The calibration of the certain soil testing equipment, such as balances, drying oven, and proving rings, should be checked periodically (annually renewed). There is always a possibility that an accident may occur during or while performing the test in the laboratory and/or at the construction field. Proper care must be taken to prevent such accident from happening by following the given of instruction manuals.

In general, the soil samples took from the desired sampling location can be group as disturbed and undisturbed soil samples. Therefore, the type of soil test equipment is somehow determined by this 2 class, plus the location of these equipments would be different depending of its functionality and effects. The effects upon its functionality attribute would normally tell how the laboratory setup would be especially involving laboratory equipments that vibrate strongly and non-vibrate apparatus. In most cases, these 2 types of soil testing equipment cannot be in the same room as it will affect the soil testing results.

Soil testing equipment would be group together in smaller section such as soil in-situ and sampling test, soil index properties, soil compaction, soil consolidation, soil bearing capacity, soil permeability, soil hydrometer test, soil vane test, and so forth. Even so, there are materials testing lab of soil which will be using the similar equipments in almost all of the civil construction projects.

Among the most used soil test equipment are weight balances, drying ovens, and ancillary items, which are required for almost all tests. Supplementary items like measuring instruments are used for measurements of length, volume, mass, time and temperature. Additionally, other major items of equipment that included into testing are such as bottle shakers, sieve shakers, centrifuge, desiccators, constant-temperature bath, vacuum, water purifier, waxing waxpot, soil mixer, de-airing water tank, riffle box, and muffle furnace. These are the common soil testing equipment that normally used in the laboratory. Knowledge of handling each of the apparatus for soil testing is essential and taking care of it is certainly recommended.

Soil Testing – Vital for Engineering Construction

The oldest raw material known to mankind that had been used for construction is soils. Therefore, in the sense of geotechnical side, soil is one of the most important and widely used engineering materials. The study of soil especially through soil testing, its physical characteristics can be determined. Soil testing for construction project is vital for its success and safety guaranteed. The analysis methods that were applied enable these properties to be used to predict its likely behavior while under defined working conditions.

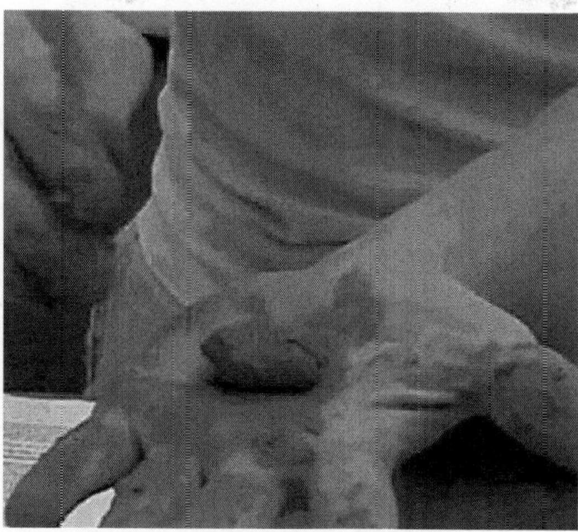

Soil mechanics is the youngest civil engineering discipline which covers the classification, description, investigation, testing and analysis of soils. Plus, it will determine its inter-relation with structures built in or upon it, or built with it whereby resulting from the soil sample testing. Soils are naturally occurring materials, which more often than not be used in their natural condition. It is unlike the other engineering materials like concrete and metals, over which control can be exercised during the manufacturing period.

Purpose of Soil Testing

Soil testing program is an analysis of the soil physical (chemical or biological, if applicable) properties. Soil sample testing would also evaluate the soil nutrient-supplying capacity at the time of sampling.

In general, there are 4 activities involved in soil testing for construction project:

- The soil sampling, as part of soil investigation process.
- The soil sample testing which normally analyze at the soil testing laboratory.
- The outcome of the analysis results obtained through soil testing. Then, interpreting the results before taking it as the safe guidelines – based on the technical standard.
- Making recommendations for the geotechnical engineers, and also the construction staffs.

For every construction site, the designated location of soil is where the soil samples would be taken into the lab for testing. In order to determine the soil physical properties, the process of soil testing would be carried out in the laboratory.

The soil sample testing can be divided into 2 main categories:

- The soil classification tests whereby indicating soil types in general form and finding the right engineering category to which it fits in.
- The engineering properties assessment of soil, for examples the compressibility, shear strength, and permeability.

Soil testing in the laboratory would produce all the required parameters, and recorded together with descriptive data relating to the soil. Therefore, the data are required by the soil engineers (geotechnical) for many construction purposes.

The more usual applications are as follows:

- The soil strata could be indentified through the data acquired from classification tests. The process known as site investigation would start exploring the subsurface conditions of the project site.
- Apart from that, the other test data would facilitate the soils engineering properties to be qualified in numerical terms. The properties of soils can then be used as the basis of analysis that normally based on the recommendations of the site investigation report.
- For ensuring the quality control assurance, the laboratory soil testing should be declared as part of the construction control measures. The design criteria should be met if comply with the lab results, especially during the construction of earthworks or excavations.
- The soil criteria of the acceptance that was used in civil construction can be drawn up in the light of available test results – this is possibly after the processing operation.
- Whenever a new ground is being opened up for construction, the findings of the site investigation can be supplemented by further testing as construction work proceeds.
- The soil sample testing data can be used for the conformation of assumptions which normally based on the previous construction experience and engineering judgement.

Types of Soil Testing

Technically, the analysis of soil samples can be done in 2 places:

- For field test – Boring test, Standard Penetration Test, Cone Penetration Test, Vane test, Field Density Test, and etc.

- For laboratory test – Moisture content, California Bearing Ratio test, Sieve Analysis test, Liquid Limit test, Compaction test, Consolidation test, Direct Shear test, Permeability test, Triaxial test, and etc.

Nonetheless, the soil laboratory testing also sightly differ as the soil samples taken from the designated place can be classified as disturbed and undisturbed samples.

Oils, Geology and Seismology

Applicability

This analysis will address constructibility issues - soils, bedrock location, topography and slopes, groundwater depth, and other geotechnical conditions, as documented in existing literature and, to the degree necessary, as investigated on-site and along interconnection routes. The Siting Board regulations also explicitly require an evaluation of geology and seismology. 16 NYCRR 1001.3(b)1(v). Furthermore, the Public Service Law requires that an Application contain, "as appropriate, geological... tsunami, [and] seismic" data. PSL164.1(a). The potential for active seismological faults and earthquakes that could cause ground motion, liquefaction, slope instability and deformation will therefore be addressed in the Application. The analyses to be performed are described below as well as in Stipulation 8.

Geotechnical Analysis

Existing Conditions: Soils on-site are a result of fill activity in the nineteenth century. Their suitability for supporting building foundations is considered poor. Piles are anticipated to be a potential method of support for the structure of the proposed Project. Bedrock is not likely to be encountered in any of the on-site excavations. The site is located in an area of the East River marked by soft soils and bedrock that is much deeper than along most of the East River. Grading and stability of slopes, other than those in temporary construction cuts and fills, should not be a concern because the site slopes very gently from the East River up toward Kent Avenue.

Further Information Requirements and Methodology

A further geotechnical site investigation will be conducted as necessary to characterize subsurface conditions in the immediate vicinity of the site for foundation design and construction, and may be reported upon in the Application. The geotechnical site investigation may consist of drilling, soil sampling, field observations, laboratory tests, and analyses that address foundation type, bearing capacity, settlement amount and rate, vibrations, liquefaction potential, seismic effects, and subgrade improvement. This information will assure that the Project is safely and adequately designed and constructed, given site-specific conditions.

The Application will include a report on the suitability of the soils actually found on-site through new or previously reported geotechnical investigation, including the suitability for construction purposes of each soil type, a description of its recharge/infiltration capacity, and a discussion of any dewatering that may be necessary during construction. Included in the Application will be a map delineating depth to bedrock and showing existing and proposed site contours.

Construction Techniques for Project and Interconnections

Steam Interconnection: The Application will assess alternative methods of steam interconnection construction, including directional drilling and tunnel boring. In addition, the possible use of the existing MTA tunnel will be discussed. The geotechnical composition below the East River will be presented on the basis of available literature. A conceptual work plan will be set forth, with descriptions of steam interconnection requirements, description of equipment to be used, impacts to bedrock geology, and contingency plans associated with directional drilling and/or tunnel boring.

Foundation Support

On the basis of the geotechnical investigation, the Application will identify the method of foundation support, and will provide sufficient technical details to establish the viability of the preferred support method.

Blasting and Pile Driving Analyses

TGE will examine the need for blasting associated with the construction at the Project site or interconnections. Although blasting is not anticipated, TGE will present a blasting plan. The plan will identify blasting operator qualifications, standard protocols, timing of discrete blasts, and what steps will be taken to ensure that any damage to adjacent structures can be avoided, or if it occurs - identified, mitigated, and compensated.

The Application will also examine pile driving, which may be used at the Project site. The Application will therefore include a preliminary plan describing the hours of pile driving operations and measures to ensure that neighboring buildings are not impacted by any Project-related pile driving. Included will be an identification and evaluation of reasonable mitigation measures regarding pile driving impacts, including the use of alternative technologies and/or location of structures, and including a plan for securing compensation for any damages that may occur due to pile driving.

Bedrock Faults and Seismology

The Application will report information on the bedrock geology that underlies the site and all interconnections. The bedrock foundations of New York City consist of a sequence of dense and stable crystalline rocks consisting of schist, gneiss, and marble that outcrop mainly in Manhattan and the Bronx. These rocks are overlaid unconformably in Queens by unconsolidated Cretaceous deposits and unconsolidated Pleistocene glacial and scattered postglacial material. The crystalline rocks of the Bronx, Manhattan, and eastern and northern Queens and Brooklyn provide the supporting foundation for major buildings and structures throughout the New York City region. These rock formations occur in two major units separated by a regional northeast-striking, eastward dipping thrust fault, known as Cameron's Line.

The Application will address the seismic risk for the Project site and interconnections. Earthquakes with Richter Magnitude greater than 6 are possible in New York City, although none are documented in the historic record. The probability of these significant earthquakes is low (about once every few hundred years), but can result in multi-billion dollar losses in a densely populated area such as New York City. To reduce the risk of earthquake damage in New York State, numerous seismic hazard reduction measures have become effective, including Seismic Building Codes. The New York State Building Code divides the state into four seismic zones, A, B, C, and D, with seismic zone factors of Z = 0.09, 0.12, 0.15, and 0.18, respectively (measuring effective peak acceleration in fractions of g, where g equals the earth's gravity acceleration).

The seismic building code for New York State uses these four zone factors, which are based of an exceedance probability of 10% in 100 years. Brooklyn is in seismic zone C, indicating a region of intermediate seismic hazard. Per the Public Service Law, tsunamis will also be addressed.

Coseismic Ground Motion, Water-level Change and Liquefaction in Sedimentary Basins

Bounded by active faults, sedimentary basins around the world are often shaken by severe earthquakes that led to some of the worst loss in human life and property damage. Liquefaction, manifested by a loss of strength and stiffness as a result of rising pore pressure in soils and sediments, commonly occurs during or immediately following earthquakes and is one of the most destructive seismic hazards in sedimentary basins. As a consequence, the assessment of the liquefaction potential of sedimentary basins has always been a key task in any earthquake-hazard mitigation programs.

Fundamental understanding of the relationship between dynamic loading, pore-pressure change and liquefaction has been largely based upon laboratory studies since the 1960's on sediments and soils subjected to cyclic loading. Significant differences, however, exist between the field and the laboratory conditions, both in the soil condition and in the loading and boundary conditions. Thus a field-based relationship between seismic loading, pore pressure change and liquefaction in sedimentary basins would be useful to improve our understanding of the seismic hazards in sedimentary basins and to verify the field applicability of laboratory results. A critical obstacle to advance in this direction has been the scarcity of adequate field data for both the ground response and the water-level change in the same sedimentary basin during the same earthquake.

Holzer et al. measured coseismic ground motion and increase of pore-pressure at a site in the Imperial Valley, California, during the 1987 Superstition Hills earthquake. Basin-wide study of coseismic ground motion, pore-pressure increase and liquefaction, however, has been lacking. The 1999 Chi-Chi earthquake in Taiwan created a watershed of data for the study of these processes, as reported here.

Taiwan is a young mountain belt formed by the oblique collision between the Luzon arc on the Philippine Sea plate and the continental margin of the Eurasian plate since 5 Ma. The 1999 Chi-Chi (M_w=7.5) earthquake, the largest to hit Taiwan in the last century, ruptured the crust along a ~80 km segment of the Chelungpu fault. Widespread liquefaction occurred during the earthquake in two sedimentary basins near the ruptured fault, causing extensive property damage. The occurrence of liquefaction is not particularly surprising in view of the magnitude of the earthquake and the geology of the affected area (Holocene sedimentary basins). However, what is noteworthy about the liquefaction in this earthquake is the enormous data resource available that may be useful for investigating the mechanisms of pore-pressure increase and liquefaction in sedimentary basins under seismic loading. A network of 60 broadband strong-motion stations in the study area recovered an unprecedented amount of near-field ground-motion data for the Chi-Chi earthquake. At the same time, 70 evenly distributed hydrologic stations, with 188 monitoring wells and hourly digital recording of the water-level at a precision of 0.1 cm, documented widespread coseismic changes in water-level across a large sedimentary basin in the Chi-Chi earthquake. The close proximity to a large earthquake, the dense distribution of the hydrological and strong-motion stations, and the availability of high-quality seismic, pore-pressure and well-log data *in the same sedimentary basin*, provided a unique and rare opportunity for studying how seismic loading affects pore pressure and causes liquefaction. The synthesis of these data has revealed a field-based, basin-wide relationship between these processes.

Spectral Velocity and Liquefaction

Sixty strong-motion seismometers in the study area functioned normally during the Chi-Chi earthquake and captured the coseismic ground response. The broadband characteristics (nominal DC to 50 *Hz*) and the high sampling rate (200 *Hz*) make the recorded acceleration traces ready for numerical integration into particle velocity and displacement time-histories, once the effect of instrument drift due to ground motion and the ensuing numerical problems are dealt with. The most commonly used parameter is the spectral acceleration (S_a), obtained at discrete periods, which is the maximum of the response acceleration of a single-degree-of-freedom damped harmonic-oscillator. The corresponding spectral velocity (S_v) and displacement (S_d) responses may also be determined from these records. Boore showed that the response spectra are largely unaffected by the baseline corrections for periods less than ~20 s.

Liquefaction in the form of sand blow, mud volcanoe and differential settlement during and immediately following the Chi-Chi earthquake were reported at numerous locations in two Holocene sedimentary basins (i.e., the Choshui River fan and the Taichung Basin) bounding the Pakuashan anticline. Since these basins are densely populated, and since the local population responded closely to the call by the Central Geological Survey immediately following the earthquake to report liquefaction occurrences, the published occurrences are likely to be fairly complete. The distribution of the liquefaction sites was highly uneven. Most of the 75 reported liquefaction sites occurred within a distance of ~30 km from the ruptured Chelungpu fault north of the Choshui River. Three sites occurring at greater distances were along the coast in artificial fills. No liquefaction occurred on the uplifted Pleistocene ridges where rocks are exposed.

In search of a relationship between the coseismic liquefaction occurrence and seismic loading, we compiled the distributions of S_a

and S_v over a wide range of seismic frequencies. The distribution of S_v at 1 *Hz* with zero damping is plotted, together with the distribution of the liquefaction sites. As expected, S_v generally decreased in magnitude with distance from the ruptured fault; but its distribution was complex in details. Comparison of this distribution with the liquefaction occurrence shows a strong correlation between the two: Among a total of 75 reported liquefaction sites, 71 sites occurred in the areas where $S_v \geq$ ~2.8 m/s. At frequencies far from ~1 *Hz*, the spatial distribution of S_v has distinctly different patterns (not shown) that do not exhibit any relationship with the liquefaction occurrence, suggesting that the sediment response to seismic loading may be frequency-dependent and most sensitive to loading at ~1 *Hz*.

Coseismic Water-level Change and Liquefaction

The hydrologic monitoring stations are located on the Choshui River fan - a Holocene alluvial fan on the west of the uplifted Pleistocene Pakuashan anticline. More than 50 stations registered stepwise co-seismic water-level changes (C_w) in different aquifers that showed no correlation with earth tides, precipitation or changes in barometric pressure. In the upper aquifer, the focus of this study, C_w was less than 0.5 m over most part of the Choshui River fan; but in an area of ~100 km^2 west of the Pakuashan anticline, C_w exceeded 3 *m*. No monitoring wells were installed in the Taichung Basin on the east of the Pakuashan anticline; thus no C_w data is available in this basin.

The occurrence of liquefaction on the Choshui River fan is closely correlated with elevated C_w in the upper aquifer: Out of a total of 13 reported liquefaction sites on the Choshui River fan, 9 sites occurred in the ~100 km^2 area where C_w > 3 *m* in the upper aquifer. This corresponds to ~10^{-1} site/km^2, as compared with an average of ~10^{-3} site/km^2 over the rest of the Choishui River fan. Since no C_w data is available for the Taichung Basin, a similar correlation cannot be made. The patterns of C_w in the lower aquifers were discussed in earlier studies. They were distinctly different from that in the upper aquifer and showed no correlation with the occurrence of liquefaction, suggesting that liquefaction occurred only in the upper aquifer. This is consistent with the fact that sediments in the upper aquifer are the youngest and least compacted, and thus most responsive to coseismic consolidation and liquefaction.

An important question is how the initial condition at the hydrological stations might have affected the above observations. We plot in Figure 3 the distribution of pore pressure 1 hour prior to the earthquake at a common depth (say, 10 m beneath the surface), normalized by the sediment overburden at that depth. Comparison between this and the distribution of the liquefaction sites shows no correlation between the liquefaction sites and the initial pore pressure. Thus we can rule out the effect of the initial condition on liquefaction in the region; and the localization of liquefaction must therefore be due to localized seismic shaking and the coseismic pore-pressure increase.

Coseismic Water-level Change and Seismic Energy

Since both C_w and S_v show a spatial correlation with liquefaction, we may expect a spatial correlation between C_w and S_v. To test this hypothesis we interpolated the S_v-values at the 20 hydrologic stations that lie within the spatial coverage of the strong-motion stations on the Choshui River fan. The interpolated S_v is plotted against the corresponding C_w in the upper aquifer. Since C_w is dependent on several other factors, i.e., the spatial variations in the permeability and the thickness of the upper aquifer, and the thickness of the confining layer above the upper aquifer, which are not represented in this diagram, we may expect some scatter in the data points on this diagram. In spite of the scatter, however, a relationship between C_w and S_v appears evident in this diagram.

The $C_w \sim S_v$ relationship in Figure 4 suggests that there is a lower threshold in S_v, below which the coseismic water level was insignificant. We choose this threshold to be 1.5 + 0.5 *m/s*, acknowledging the fact that the scatter in the data prevents a clear choice; i.e.,

$$C_w \sim 0, \text{ if } S_v \leq 1.5 + 0.5\ m/s.$$

Above this threshold, we fit the data with a simple function:

$$C_w\,(m) = 1.74\,(S_v - 1.5)^2,$$

The goodness of this fit is given by the square of the correlation coefficient, 0.97; the standard error is 0.1 *m*.

The correlation between the coseismic liquefaction with high S_v may be expected because S_v is related to the seismic energy per unit mass ($S_v^2/2$) that supplies the required energy to break up the bonds between the sediment grains and to allow relative movements among grains - promoting coseismic consolidation, pore-pressure increase and liquefaction. What is noteworthy about Figure 2a is that it provides for the first time a field-based, basin-wide relation between coseismic liquefaction and elevated S_v at ~1 *Hz*. Coupled with recent advances in strong-motion modeling, this relation may provide an additional means for predicting coseismic pore-pressure increase and liquefaction in sedimentary basins.

The correlation between the coseismic liquefaction and elevated C_w may also be expected because rising pore pressure in sediments under seismic loading is widely believed to be the precursor for the coseismic liquefaction. It should be noted, however, that pore pressure in the aquifers is not always equal to the water pressure in wells and that liquefaction may not occur at the same depth as that where water pressure was measured. Thus the coseismic increase in pore pressure in the liquefied layer may not be the same as the recorded value. There is good reason to believe, however, that the recorded water level at the Yuanlin station may indeed represent the coseismic pore-pressure increase in the liquefied layer: At this station, two wells were drilled to the upper aquifer, but screened at different depths; in spite of the depth difference, the recorded coseismic water-level changes in the two wells were identical (6.55 m). The depth of liquefaction may be estimated on the basis of the effective-stress principle that the effective stress at the liquefaction depth should be equal to the increase in pore pressure. Thus a coseismic water-level increase at the Yuanlin station implies an effective stress of 6550 *Pa* and a liquefaction depth of 4 - 7 *m*, depending on the thickness of the saturated column above the liqefied layer. This estimate of liquefaction depth is in substantial agreement with that determined from boring at the liquefied site by Hwang and Yang shortly following the Chi-Chi earthquake.

Holzer et al. made simultaneous measurements of the ground acceleration and coseismic pore-pressure change at a single site in southern California, which was undergoing liquefaction during the 1987 Superstition Hills earthquake.

They showed that pore pressure did not change until the peak ground acceleration (PGA) reached 0.21 *g*. To compare this result with the pore-pressure change on the Choshui River fan during the Chi-Chi earthquake, we determine the PGA from the acceleration records and plot the interpolated PGA at the 20 hydrologic stations that lie within the spatial coverage of the strong-motion stations on the Choshui River fan against the corresponding C_w at the hydrological stations. No relationship between C_w and PGA is evident. On the other hand, plotting the interpolated S_a (1 *Hz*) against C_w reveals a strong correlation between C_w and S_a :

$$C_w = 0, \text{ if } S_a < 1.0\, g$$

$$C_w(m) = -3.02 + 1.07 \exp [S_a (g)], \text{ if } S_a > 1.0\, g$$

The goodness of this fit is given by the square of the correlation coefficient, 0.98; the standard error is 0.06 *m*.

Laboratory data for soils under cyclic loading (16) shows there is a threshold in the shear-strain amplitude below which loading does not cause a buildup of excess pore pressure, but above which increasing strain amplitude causes rapid increase in pore pressure. For different sands this threshold is on the order of 10^{-4}. Following, we estimate the dynamic strain in the Chi-Chi earthquake by dividing S_v with the average shear-wave velocity (~400 m/s) estimated for the top soft layers in the Taichung Basin, while acknowledging the uncertainty from using an average shear velocity. The thresholds in S_v for initiating pore-pressure increase and liquefaction thus imply strain thresholds, respectively, of ~4×10^{-3} and ~7×10^{-3}.

Thus, while the field-based estimate of the lower threshold in strain amplitude for initiating pore-pressure increase is more than an order of magnitude greater than that estimated from the laboratory data for different sands, there is substantial agreement between the field and the laboratory results in the strain amplitude required for initiating liquefaction. We interpret the discrepancy in the threshold for initiating pore-pressure increase as due to the different conditions of the sediments in the field and in the laboratory: While cohesionless sands are used in most laboratory experiments, the sediments in the Choshui River fan are lightly cemented by secondary minerals precipitated from groundwater, which imparted a cohesion to the sediments, which in turn may have raised the threshold in the strain amplitude required for coseismic consolidation and pore-pressure increase to occur in the field, beyond that determined in the laboratory. Once this initial resistance is overcome, sediments in the field would consolidate to cause pore-pressure increase in the same manner as that in the laboratory. At liquefaction, the sedimentary grains in the field and in the laboratory are fully mobilized such that whatever mechanism (e.g., cementation) that caused the differences between the threshold strains for initiating pore-pressure increase in the field and in the laboratory, may no longer be active at the strain amplitude that initiates the liquefaction.

The pore pressure and the particle velocity in a watershed at Sespe Creek, CA, induced by several large earthquakes. From these they estimated a threshold of 5 - 20 cm/s for the particle velocity to initiate pore-pressure increase, which is an order of magnitude lower than that estimated in this study (1.5 m/s). It is difficult to explain the difference between the two studies, but we suspect it may partly be due to the different tectonic settings in the two areas, i.e., strike-slip setting at Sespe Creek versus foreland setting in Taiwan, and partly be due to the different approaches in the two studies, i.e., model calculation in versus data analysis in this study. In spite of some scatter in the data, a field-based, basin-wide relation appears to exist between the coseismic water-level changes, ground motion and liquefaction in the Chi-Chi earthquake. The large difference between the laboratory-based and the field-based results in the lower threshold in the dynamic strain for pore-pressure increase calls for caution in transferring laboratory results to the field applications. The similarity in the strain amplitude for initiating liquefaction in the field and in the laboratory suggests that whatever mechanism(s) that caused the difference between the threshold strains that initiated pore-pressure increase in the field and in the laboratory, was no longer active at the strain amplitude that initiated the liquefaction.

Acknowledgements

This work was partly funded by NSF grants EAR-0106802 and EAR-0125548, and partly by NSC grant NSC91-2116-M-001-017. We thank Michael Manga for discussion and for calling to our attention the references listed in (11); we also thank Emily Brodsky, James Kirchner and Lorraine Wolf for discussion.

The Relationship between Soil Microbiology and the Biodynamic Preparations

Mycorrhizal fungi

There is a very interesting and easily read book for the layman on soil microbiology, which has been only recently published called "*Tales from the Underground*" by David W. Wolfe from Cornell University USA. It is published by Perseus Publishing Cambridge Massachusetts. Wolfe talks, among other things, a lot about soil bacteria and soil fungi and he describes the amazing connections between the bacterial and fungal life in the soil and plants. In particular the connections of the rhizobia bacteria (those are the bacteria that are responsible for the formation of nitrogen nodulation of the legume plants) and also the various mycorrhizal fungi which attach themselves to the roots of many plants in a symbiotic fashion. Mycorrhizal fungi form an association, with various plants, by means of their fine hyphae attaching themselves to the plant roots and seek out nutrients and moisture for the plant, in exchange for carbohydrates that the plant synthesizes by photosynthesis.

I have been looking at the nitrogen nodules on legumes resulting from nitrogen fixing bacteria, with the naked eye all my life and I am very familiar with the inoculation of various rhizobium species needed for different legumes. The rhizobia are host specific. Rhizobia needed for lucern are different from those needed for say soya bean, vetch or cow pea.

But of the mycorrhizal fungi I knew very little about except they existed on and grew around the roots of the pine trees, the silver birch and the oaks and aided their growth, but I had no idea that apparently 90% of all plants right through the world have a symbiotic relationship with the mycorrhizals. With an enlargement scope of 20X, these hyphae can be seen in and around the roots of the host plant and right through the particles of the surrounding soil. They are very fine and cobwebby looking, about 1 micron (finest merion or cashmere wool is 15-20 micron) and which penetrate the particles of soil. Not surprisingly they look like strands of fungus on mouldy bread. Incidentally these hyphae help bind the soil particles to form a crumb structure. Wolfe points out that the mycorrhizal fungi unlike the rhizobia bacteria are not always host specific and often spread from plant to plant and species to species. The strands of their hyphae are interconnected and apparently travel for very long distances seeking nutrients for the plants. Also often along with various helpful bacteria, plants are able to get their nutrition a long way from their root zone. A kind of food conveyor belt is in progress.

For instance, it is thus possible to have a symbiotic relationship between, the rhizobium bacteria which fix nitrogen from the atmosphere via the nodules on the roots of legumes, which are being grown between a crop of maize or sugar cane, and mycorrhizal fungi whose hyphae helpfully pass on the nitrogen from the legume to the roots of the crop. Apparently this is quite a comparatively recent discovery in research into the world of the mycorrhizal fungi and only since the last 10 to 15 years.

The Agriculture Course

Rudolf Steiner while giving the agricultural course in 1924 talked about the preparations for the first time and how they would enrich the manures and composts and also bring a sensitivity to the plant. When he had just finished talking about the dandelion preparation BD506, he said (in Chapter 5, page 104 Agricultural Course), “Now if you treat the soil as I have described, the plants will be able to draw on what they need from a very wide area. They will be able to use not only what is in their own field, but also what is in the soil of a nearby meadow if they need it or what is in the soil in a neighbouring forest.” He was obviously talking about the connecting hyphae of the mycorrhizal fungi.

So previously when I read this extract of Rudolf Steiner's lecture, I had always found it difficult to understand. Now in relation to Wolfe's explanation of the activity of the hyphae and the distances they could travel to collect nutrients and moisture, it all suddenly made sense! We are using the preparations, in one way to encourage this microbial soil life and the resultant balanced soil which supports healthy plants. Thus one function of the preparations is to improve the effectiveness of these soil mycorrhizae. And another function is that the healthy living soil allow the cosmic growth forces to stream into the soil to support plants of healthy and nutritious quality.

Research Work

There has been some work done recently by Prof. Barbara von Wechmar, a microbiologist at the Stellenbosch University in Capetown, So. Africa, of growing fungi and bacteria samples from the BD preparations and CPP in the laboratory on various substrates. These all showed a tremendous variety of fungi and bacteria. The BD500 picture was described by Prof. Barbara as an interesting and beautifully balanced collection of fungi.

Dr. Perumal from the MCRC Institute in Chennai, Tamil Nadu, India has tested the BD preparations for bacteria and particularly rhizobia. He found the BD500 cow horn dung and the BD504 Stinging Nettle have remarkably high numbers of this bacteria.

Work done at ICRISAT in Hyderabad, Andhra Pradesh, India has found beneficial bacteria, antagonistic towards Fusarium root fungus of chick peas, in compost treated with biodynamic preparations, which effectively combated the fungus.

This shows the benefits of the beneficial bacteria in the BD preps and which can be multiplied during the composting process.

The Preparations

The Biodynamic preparations, when applied to the soil in the various ways – through the stirred cow horn dung BD500, and the preparations BD502-507 which are used in composts, liquid manures or cow pat pit (CPP) – on one level actually work through the soil micro life and encourage the development of mycorrhizal fungi hyphae and the rhizobia as well as other micro and macaro soil organisms. They thus work to enliven the soil structure and make the plant nutrients more available. Again to quote Rudolf Steiner, "The plants then will be able to use not only what is in their own field, but what is in the soil of a nearby meadow if they happen to need it or of what is in the soil in the neighbouring forest." The forces that are in the preparations work through the microbial life in the soil or the healthy nature of the soil. Cosmic influences come into the soil through healthy living soil.

Actually the preparations make the organic part of the soil active and living. *The biodynamic farmer can actually see the changes in the structure of his soil* and also most importantly the health and quality of his crops.

The enlivening of the soil through the increase of the living organisms therein, is the most important aspect of the biodynamic system of agriculture. Actually it is the preparations which make organic farming work.

We can now come to the thought that there is a spiritual impulse or connection, working out there between the soil, plant and the cosmos. The preparations when used on the land, can be seen as a bridge between the life of the soil and the wholesome growth of all plants. Rudolf Steiner talks about nature spirits or elemental beings which are also a bridge between the world of spirit and the world of plant and the world of animals. Thus, it is great if the farmer can develop a connection with his plants and his animals with enthusiasm for their well-being.

Rudolf Steiner calls it the etheric formative force, where Spirit is working out of Matter.

Soil is a Habitat for Living Organisms

What does soil look like from the inside?

Using your imagination is an important part of developing an appreciation of the environment that organisms living in the soil experience. If you cannot imagine what organisms experience in soil, it is difficult to appreciate the many ways in which they are influenced by their surroundings.

Novel techniques are available for looking directly into a soil to observe living organisms. For example, special video cameras can be used with time-lapse cameras to follow the growth of fungi or the movement of animals.

Borescopes and mini-rhizotrons are two tools that can be used in this way. This technology provides an image of the habitat of soil organisms.

Why is Soil Such a Diverse Habitat?

The physical and chemical characteristics of a soil are different in different parts of the soil profile. Generally, the upper layers of a soil have more organic matter and roots than the lower layers. Other differences are related to the nature of the soil constituents, to weathering processes and to past land management practices.

The origin of the soil determines the particle size distribution, which in turn affects the way the soil is packed, creating spaces and surfaces that are either accessible or inaccessible to soil organisms. The proportion of clay or sand influences the structural characteristics of soil and determines its response to management practices.

The structure of a soil is related to the extent to which particles are aggregated into relatively stable formations. Aggregates vary in size and the degree of soil aggregation dictates the number and size of the pores within it.

The type and quantity of organic matter also influences the degree of soil aggregation. Soil structure determines the number and size of air spaces and influences the water holding capacity and drainage properties of the soil.

Roots can also alter the soil environment for living organisms and play an important role in supplying nutrients for organisms that live around them. Roots are dynamic structures, especially when young, releasing carbon compounds that can be readily used as an energy and carbon source by many soil organisms. However, over time roots change, and older roots have different surface properties and release different compounds to younger roots. As roots age, the outer layers of the root die, providing a source of organic material as well as a habitat for soil organisms. It is important to consider the soil as a continually changing and complex environment for organisms. Part of this complexity arises because the organisms themselves contribute to the some of the changes that occur within the soil.

What Environments do Organisms Experience in the Soil?

The habitat available to a soil organism depends on its size: the soil environment of bacteria will be quite different to that of an earthworm. For a bacterium, important aspects of the soil environment include:

Microhabitat (structure):

- surfaces of soil particles;
- pore spaces;
- roots;
- dead organic matter (plant, animal or microbial);
- water films

Physical and chemical characteristics:

- the charge of soil particles (positive or negative);
- degree of aeration;
- pH;
- salinity;
- temperature;
- nutrients.

Biological aspects:

- other living organisms (including roots).

Bacteria are usually attached to the surfaces within soil pores or fragments of organic matter. They may attach themselves using flagella or fine hair-like fribrillae, although not all bacteria have these structures. Another method of attachment occurs through the production of polysaccharide gums that are released through the cell wall of some bacteria. Certain fungi also produce gums that actually help bacteria attach to soil particles.

Fungi experience a similar soil environment to that of bacteria, but the scale is greater. Fungi spread much further through the soil than do bacteria and will encounter a greater variety of soil environments. Also, bacteria have access to smaller pore spaces than most fungi. Protozoa and mites both feed on bacteria but only the protozoa are small enough to enter some soil pores. Mesofauna such as mites and collembola cannot enter smaller soil pores and therefore bacteria that occur there are protected. The smallest soil pores are inaccessible to bacteria and may contain organic matter that is protected from degradation.

While the soil environment obviously has a great effect on soil organisms they, in turn, may affect their physical environment. The breakdown of organic matter by soil organisms changes the structure of the soil, which leads to changes in the habitats available for soil organisms. The chemical environment of soil organisms can also be of their own making. Some bacteria and fungi initiate many of the biochemical reactions that occur in soil. The decomposition of leaves requires the enzyme cellulase, which is produced by soil organisms. The activity of cellulase changes with time, with the main peak in activity occurring soon after the addition organic matter, such as red maple leaves, to soil (Linkins *et al.* 1999).

This reflects the increased production of cellulase by soil organisms in response to organic matter. Cellulase enzymes occur within the organisms (endocellulase), but can also be excreted into the soil (exocellulase). Thus, the data reflect the greater cellulase activity during the early stages of decomposition.

Waste products excreted by soil organisms provide substances that can be degraded further by other organisms. Alternatively, the wastes may be toxic to some organisms and inhibit their growth.

Some organisms tolerate a wide range of conditions whereas others are adapted to a more limited environment. For example, cyanobacteria are rarer in acidic soils than in soils of pH 7 to 8. The abundance of bacteria is less in acidic soils than in soils of higher pH. Generally, there is not such a strong relationship between soil pH and fungi as there is between soil pH and bacteria, although individual fungi can have a marked pH preference.

Soil habitats differ greatly depending on land use. For a similar soil type within the same climate zone, a forest soil will generally have a greater diversity of habitats for soil organisms than a cultivated agricultural soil. These differences are primarily associated with a greater diversity of plant species and heterogeneity of the soil itself and the characteristics of organic matter produced in natural and agricultural ecosystems. The size of soil pores, and therefore the habitat available to bacteria etc, depends on soil structure. For example, soils with high levels of clay have a greater proportion of small pores less than 0.2 μm in diameter than loams on sandy soils (Hassink *et al.* 1993). The very small pores are too small for bacteria to enter, providing protection to organic matter.

Soils with a higher number of small, but not too small, pores (e.g. 0.2 to 6 μm in diameter) have more bacteria than do those with fewer pores in this size category. This is because bacteria survive best within aggregates of soil rather than on the surfaces of soil particles outside aggregates. Bacteria exposed on surfaces are likely to be eaten by soil animals such as mites and springtails (collembola) or damaged by extreme cycles of wetting and drying. Therefore, a clayey soil would be expected to have a greater number of bacteria than a sandy soil under similar land use. For one of the soils studied by Hassink *et al.* (1993), there was a close relationship between the biomass of bacteria and the percentage of the pore volume in soil that was in the size class 0.2 to 1.2 μm.

Bacteria and fungi commonly occur within aggregates of soil, although they are not distributed evenly in all size fractions of soil aggregates. Most bacteria in this study were present in soil aggregates that ranged from 2 to 20 μm in diameter.

The number of soil organisms varies greatly between the surface and very deep layers in the soil profile. The main reason for this is because the supply of plant organic matter that is essential for many soil organisms is almost absent lower in the soil profile. However, it needs to be remembered that organisms do occur at great depths within the regolith, even though their numbers are much reduced compared with communities at the soil surface.

How does Land Use Alter the Habitat of Soil Organisms?

Soil disturbance has major effects on soil habitat diversity. Disturbance changes the physical uniformity of soil and in natural ecosystems, it changes the diversity of plant species growing in the soil. Disturbances are not the same and they can have different impacts in different soil types.

Two examples of the effects of disturbance in different soil environments are:

- severe soil disturbance due to cultivation tends to create a more homogeneous soil environment than does disturbance of a forest during a process such as logging.
- the higher diversity of plant species in a natural forest creates more habitats in soil than does a more uniform plant community (such as a crop grown in monoculture).

When soil conditions are altered in any way, the number of organisms changes and the relative abundance of different types of organisms in the soil changes as well. A change in environment may favour one group of organisms so that its abundance increases, while the number of other organisms decreases either because the conditions are less favourable or because of the increased influence the organisms that have become dominant. An example of this is the change in soil organisms following the addition of nitrogen and straw to soil (Russell 1973). Adding nitrogen increased the length of hyphae, the number of bacteria and the number of amoebae. The thickness of the hyphae also increased when nitrogen was added. The change in thickness of hyphae is likely to have occurred because the increased nitrogen favours species of fungi with thicker hyphae.

Liming a soil by the addition of calcium carbonate can also change the number of bacteria and fungi present (Russell 1973), although the response is not always predictable. In this case, the outcome depended on the soil characteristics and the types of organisms present. In one soil, the addition of calcium carbonate increased the number of bacteria but decreased the abundance of fungi. In contrast, the addition of calcium carbonate to a second soil decreased the abundance of bacteria and had little effect on the abundance of fungi.

Do Organisms Live Deep in the Regolith?

Living organisms are not restricted to the surface layers of soil. Bacteria are the most common organisms at depth, and they even occur at more than 1 kilometre below the earth's surface. These bacteria cannot rely on sources of plant organic matter to meet their need for carbon and energy. One possible source of carbon for organisms deep in the regolith in some regions is oil. Microbial activity in oil wells, including their role in the corrosion of drilling rigs, has been well documented indicating that they are able to live and thrive using oil as an energy source.

Bacteria living at a great depth in the regolith have to be able to survive in environments with low levels of all nutrients. But they appear to be uniquely adapted: they are small (most are less than 1 μm in size), generally have low rates of respiration and can exist in a physiologically stressed state.

We know relatively little about the biology of organisms that occur deep in the soil. It is difficult to study deep soil organisms. This is because their surroundings are often greatly altered during the investigation from what they normally experience and few deep soil bacteria will grow on artificial media. For example, deep soil organisms may experience increased or even toxic levels of oxygen when they are brought into the laboratory.

To gain accurate information about the biology of organisms that occur deep down in the soil the experimental conditions need to resemble as closely as possible the conditions normally experienced by such organisms. If these requirements are not met, the experimental results may not represent what actually occurs at depth in the regolith. This is important for studies of all soil organisms and explains why investigations of organisms in artificial media need to be scrutinised carefully before conclusions can be drawn about the function of the same organisms in field soil.

Chapter 4

Organic Matter of the Soil

Organic matter is one of the main components of the soil and conditions its fertility. According to their composition, the organic compounds of the soil are unique and complex. They are formed from plant and animal residues as a result of microbiological metabolism.

All organisms living above the earth's surface and in the soil (animals, plants, and microorganisms) find their way after death to the soil, where they are metabolized by the living cells of microbes which form various substances. These substances, in their turn, are subject to biochemical transformations, as a result of which specific, relatively stable, and complex compounds called humus are formed.

Higher plants supply the soil with organic compounds during the period of vegetation, releasing various nitrogenous, or nonnitrogenous compounds from their roots. They also shed dead fractions of roots and parts growing above the surface. The total mass of plant residues entering the soil may reach considerable proportions. For example, in forests, the annual fall of leaves and twigs comprises 1.5-7 tons/hectare according to the type of the forest, its age, and the climate an soil conditions. Various woods yield different amounts of residue. Annual fall of leaves and twigs according to Zonn (1954), is as follows (average figures):

Deciduous forests 2.7 ton/hectare;

Oak forests 3.9 ton/hectare;

Pinewood forests 4.1 ton/hectare;

Fir tree forests 6.0 ton/hectare.

Thus, fir-tree forests rank first according to the amount of leaves and twigs shed, with pinewood, oak trees, and deciduous forests following.

The amount of the forest litter formed varies. The largest amount is to be found in fir-tree forests (50 tons/hectare and more).

According to the data given, the amount of organic compounds in soils of various kinds of forests varies. The amount of organic compounds shed in fir-tree forests reaches 5.85 tons/hectare; in pinewood forests, 3.96 tons/ hectare; and in oak forests, 3.5 tons/hectare (Zonn, 1954),

As should be expected, the "sheds" of different forests differ qualitatively, too. According to the data of Zonn, the fall of fir trees is more acid than that of pine or oak trees. According to our observations, the leaves of birch and lime trees in June are decomposed in the soil more rapidly than the leaves of oak, aspen, or the needles of pine trees.

Meadow vegetation yields dry mass (from the parts growing above the surface) amounting to about 2-6 tons/hectare and roots, 7-11 tons/ hectare. In the chernozem meadows of the steppes about 7 tons/hectare of dry mass (parts grown above the surface) were found and 25 tons of roots. In steppes on solonets soils 5 tons/hectare of the dry mass of the parts growing above the surface and 13 tons of roots were found (Savvinov and Pankova, 1942). In the desert steppes on serozem, about 1 ton/hectare of the mass growing above the surface and 15 tons of roots were found (Kul'tiasov, 1925). According to Kononova (1951), grass yields about 21 tons/hectare of root mass and, according to Belyakova (1953), the weight of roots of lucerne reaches 40 tons/hectare. Annual grasses yield less root mass than perennials (Vilenskii, 1954).

Plant tissues are composed of various carbon and nitrogen compounds. They contain sugars, dextrins, starch, pectic and tannic substances, organic acids. fats, waxes, tars, and many other compounds.

The main component of the plant material is cellulose $(C_6H_{10}O_5)n$. It constitutes the cell wall. Cellulose comprises 85-90% of the total weight of cottonseed fibers, and about 50% of bark.

The cellulose is decomposed by special cellulose microbes—bacteria, myxobacteria, actinomycetes, and fungi. Various intermediate compounds are formed in the decomposition process: organic acids, alcohols, sugars, and others.

Hemicellulose, in addition to cellulose, also appears in plant cells. Hemicellulose is easily hydrolyzed by acids and alkalis with the formation of sugars, uronic acids, and other compounds. In wood, cellulose is impregnated with lignin, the content of which reaches 34%. Lignin differs from cellulose by its higher content of carbon (62-69%, in cellulose only 49.4 %) and lower content of oxygen. Upon oxidation it yields aromatic compounds. The chemical structure of lignin has not been ascertained. Lignin in the soil is decomposed by microbes with the formation of final decomposition products, CO_2 and water, or intermediate products.

Proteins are the most common nitrogenous compounds present in the cells of plants, animals, and microbes. They are present in protoplasm, nuclei, and in various protein reserve substances (metachromatin, protein crystals, aleuron grains, etc). Complex proteins are known—proteids and proteins proper such as globulins which are insoluble in water but soluble in dilute salt solutions; water-soluble albumins; prolamins—proteins of the gluten of the wheat grain (gliadin) which are soluble in 80% alcohol; glutelins—plant proteins, soluble in dilute alkaline solutions; sclero-proteins—insoluble proteins of horny tissues such as collagen, keratin, and others.

Many complex protein compounds are known, such as phosphoproteids containing phosphorus, nucleoproteids—proteins of the cellular nuclei and nuclear inclusions (which upon hydrolysis are decomposed into simple proteins and nucleic acids containing phosphorus), chromoproteids—proteins containing pigments (e.g., blood hemoglobin and some antibiotics formed by microbes), and glucoproteids (mucoproteins), which are proteins containing carbohydrates.

Other complex proteins which are present in plant, animal and microbial cells are: albumoses and peptones, which are protein compounds forming colloidal solutions and giving biuret reaction, and amino acids, which are colorless watersoluble compounds containing amino groups ($-NH_2$) and carboxyl groups (OH-C=O).

Plant residues as well as the dead cells of microbes and animal organisms find their way into the soil, where they are subject to physicochemical and biological processes.

The main transformations of plant residues are carried out under the influence of biotic factors. The dead parts of plants in the soil begin to decompose immediately, at first under the action of their own enzymes and then quite rapidly (perhaps simultaneously) under the action of microbial enzymes.

The first to be decomposed are the easily assimilated organic compounds: sugars, organic acids, and alcohols; then follow proteins, amino acids, fats, pectins, gums, hemicellulose, and lastly cellulose and lignin. The soil microbes also decompose waxes, tars, and many other stable compounds. It can be said that no organic compound exists which cannot be decomposed by microorganisms. Some of them are decomposed rapidly (carbohydrates, proteins, etc.), and others, slowly (tars, waxes, etc.).

The decomposition of organic compounds may be carried out to the final products, CO_2 and water, or may stop with the formation of intermediate compounds. The latter may be in the form of organic acids, alcohols, amino acids, etc.

Simultaneously with the decomposition of organic compounds, synthetic processes are taking place in the soil. The so-called autotrophs are known to synthesize organic compounds by assimilation of CO_2. The first of these are the photoautotrophic algae, which may be present in considerable numbers. Many colorless chemitrophs and pigmented bacteria possess the same capability. They assimilate carbon dioxide and synthesize organic compounds at the expense of chemical or light energy. Among these are the nitrate, sulfur, iron, hydrogen, and methane-oxidizing bacteria. The sole source of carbon for these organisms is CO_2. Their energy requirements are satisfied by the following simple compounds: ammonia, nitrates, sulfurous and ferrous compounds, hydrogen, methane, and others.

Many heterotrophic microorganisms are capable of assimilation of CO_2 and of synthesizing organic compounds. This capability was detected in the representative of the genera *Pseudomonas* and *Azotobacter,* in sporogenous and asporogenous bacteria, in yeasts, fungi, and in actinomycetes.

The synthesis of organic compounds may reach considerable dimensions, 5% and more of the CO_2 supplied during the experiment (Liener and Buchanan, 1951).

Cells dividing in the logarithmic phase of growth assimilate ten times more CO_2 than in other stages of growth (Mac Lean et al. 1951; Shaposhnikov, 1952; Rabotnova, 1950; Linsh and Calvin, 1952; Citterman and Knight, 1952, and others). According to Werkman and Wilson (1954), all microorganisms, autotrophs and heterotrophs, are endowed with the ability to assimilate CO_2, but to a different extent according to the type and conditions of culture growth.

The synthesized compounds and decomposition products of plant residues, as well as other organic compounds, find their way into the soil solution and are utilized as nutrients by microbes and plants.

Shmuk (1930) noted the presence of the following compounds in the soil: nitrogenous compounds (methylamine, choline, histidine, arginine, lysine, cytosine, xanthine), fats, organic acids (oxalic, succinic, crotonic, acrylic, benzoic, etc. esters (glycerides of caprylic and oleic acids), carbohydrates (pentoses, pentosans, hexoses, cellulose and its decomposition products), alcohols, aldehydes, tars, paraffins, and other compounds.

Davidson, Sowden, and Atkinson (1951), employing the method of paper chromatography, detected about 30 compounds in the organic fraction of the soil: such as arginine, histidine, lysine, alanine, leucine, proline, isoleucine, valine, aminovaleric acid, aspartic acid, tyrosine, threonine, glutamic acid, and others.

According to Kejima (1947), the following acids can be detected in the soil: 6-7% aspartic acid, 5% glutamic acid, and 18% of other amino acids, totaling 31.9% total nitrogen. According to the author, 66-75% of soil nitrogen is not in the humus but in microbial proteins.

Such compounds as polyuronic acids which comprise either components of plant tissue (hemicellulose) or products of microbial synthesis (slimy compounds constituting bacterial capsules) can also be detected In the soil.

Schreiner and Reed (1907) isolated various organic nitrogen and carbon products from fertile soils. Creatine, xanthine, hypoxanthine, adenine, and cysteine were among the first detected.

Rudakov and Birkel' (1949) found uronic acids among the metabolites of plant roots. The liberation of these acids takes place with the participation of bacteria possessing protopectinase.

Shori isolated allantoin from the soil and Enders obtained methyl glyoxalate, a compound which, according to Neiberg, is an intermediate in hexose fermentation and, according to Gebert, a primary structural element of protolignin. It is assumed that methyl glyoxalate is an intermediate compound, "a bridge" which links the lignin and cellulose theories of the origin of humic acids (Kononova, 1951). There are various biologically active compounds in the soil: vitamins (B_1, B_2, B_6, B_{12}), auxins, pantothenic, nicotinic, folic, and para-aminobenzoic acids, biotin, and other compounds activating the growth of plants and microbes, There are also inhibitors that may suppress the growth of plants (toxins) and microbes (antibiotics), and free enzymes—catalase, peroxidase, invertase, amylase, tyrosinase, and others.

Investigations show that enzymes exist in the soil in the active state. Their quantity varies in accordance with the soil composition, season, and climatic conditions. In fertile cultivated soils there are more enzymes than in poor nonfertile soils. The more organic compounds in the soil, the more active is the growth of microbes and the greater the enzymatic activity of the soil (Hoffmann, 1952). The upper layers contain more enzymes than the deeper ones.

The liberation of CO_2 has been observed to depend on the enzymatic activity of the soil (Seegerer, 1953; Ukhtomskaya, 1952). According to the data of Ukhtomskaya, the amount of enzymes in the soil increases proportionally to the amount of organic compounds introduced. The enzymatic activity of the soil is more pronounced in May than in October, when the microbiological processes diminish.

Table: *The enzyme content of the soilThe amount present in 100g of soil, expressed as decomposed substrate: catalase and peroxidase in ml of 0.1 N KMn0$_4$ solution; protease, in mg of nitrogen; amylase, in mg of maltose; invertase according to inversion, in mg of glucose (Ukhtomskaya, 1952)*

Enzymes	*May Control*	*May 500 tons*/ hectare*	*May 1,000 tons/ hectare*	*May 2,000 tons/ hectare*	*October Control*	*October 500 tons/ hectare*	*October 1,000 tons/ hectare*	*October 2,000 tons/ hectare*
Amylase	29	1,132	3,568	4,320	71.0	596	1,606	1,870
Invertase	29.69	428.2	2,012	2,173	87.5	333.7	712	1,182
Protease	48.0	62.68	61.25	84.0	42.5	53.2	54.2	68.48
Catalyse	279	601	671	723	260	741	470	980

*The organic compounds were introduced with sewage waters.

Kuprevich (1949) detected the presence of the following enzymes in the soil: catalase, tyrosinase, phenolase, asparaginase, urease, invertase, amylase, and protease, noting that their accumulation depends on soil cultivation. The quantitative figures for catalase, invertase, and urease present in soils, according to his data, are given.

Table: *The activity of extracellular soil enzymesCatalase in cm^3 of 0$_2$ produced in three minutes at 18i C; invertase in mg of inverted saccharose; urease in mg of decomposed urea (Kuprevich, 1949)*

Soils	*Catalyise*	*Invertase*	*Urease*
The soil of the garden of the Botanical Institute of the USSR Academy of Sciences in Leningrad	6.0	167	34
The soil of a pine forest	6.4	202	41
The soil of an orchard	7.9	220	70
Washed river sand under barley	0.4	0.0	15(?)

Sorensen noted greater activity of xylanase in cultivated soils than in noncultivated soils. The enzymatic activity increased six times and more when straw or xylan were applied to the soil (Sorensen, 1955).

Scheffer and others (1953) and Seegerer (1953) pointed out the increased activity of invertase and urease after the application of organic fertilizers, especially manure. The amount of enzymes in the soil also depends on the vegetative cover. When a green crop of serradella was plowed in, the amount of catalase and invertase was greater than after the plowing in of green lupine.

Table: *The catalase and invertase content in 2 g of soilCatalase, in cm³ O_2 from 5 ml 3% H_2O_2. Invertase, in mg of inverted saccharose*

Soil	*Catalase August*	*Catalase Sept.*	*Invertase August*	*Invertase Sept.*
Fallow (control)	4.5	4.6	17.43	6.02
After introduction of lupine	5.6	6.7	27.63	32.41
After introduction of serradella	5.3	7.3	29.34	39.67

As can be seen from the given data, the enzymatic activity is closely correlated with the activity of microorganisms. Any increase In the amount of the latter leads to enhancement of enzymatic soil processes. Hoffmann (1951) considers that the enzymatic activity of the soil is an index of its fertility.

There are data in the literature indicating that plant roots excrete various enzymes into the soil, such as catalase, tyrosinase, amylase, protease, lipase and others. All these organic compounds comprise only 10-15% (approximately) of the total organic mass of the soil. However, owing to their great activity, they are of considerable importance. Many of these organic compounds (vitamins, auxins, certain amino acids) are catalysts of biological and biochemical processes in the soil.

The part played by free extracellular enzymes is not yet clear, but they may be assumed to be important in transformations of many types of organic compounds and, in particular, in the synthesis of humus compounds. We should note the considerable role of antibiotics in the life of the soil. These substances influence the composition of the microbial populations and this affects many soil properties.

Humic substances of the soil. Humus comprises the bulk of the organic soil compounds and is responsible for the dark coloration.

Humus is a mixture of various and very complex natural compounds. The uniqueness of these compounds does not allow for their classification into any of the groups of compounds known to organic chemistry. These substances are synthesized in the soil, apparently exogenically, by the action of extracellular enzymes. The composition of humus is more complex than that of many compounds of plant and microbial organisms. Humus comprises 85-90% of the total organic matter of the soil.

The chemical composition and origin of humus is not as yet clear. Characterization and subdivision of humic soil substances is based on external features.. color, and its relation to solvents. The main components of humus are assumed to be the three acids: ulmic, humic, and crenic.

According to Vil'yams, ulmic acid is formed during the anaerobic decomposition of organic compounds by anaerobic microbes. It is easily soluble in water imparting a dark brown color. It forms water-soluble salts with monovalent cations (potassium and sodium) and insoluble salts with bi- and trivalent cations. Under the influence of external factors, such as low temperature (freezing) or drying, ulmic acid is converted to water-insoluble ulmin.

Humic acid is formed under aerobic conditions and is considered to be a product of bacterial and fungal metabolic activity. Its properties are close to those of ulmic acid. It is less soluble in water than ulmic acid and gives the soil a black color. It is also denatured and converted into an insoluble compound, humin. It forms water-soluble salts with monovalent cations and insoluble salts with biand trivalent cations.

Humic acid has been studied in more detail. The following organic groups were detected: carboxyl (COOH), hydroxyl (OH), carbonyl (CO), and methoxyl (CH_20), (Kononova, 1951).

Humus contains from 10-40% humic acids. The largest amount can be found in chernozems.

Humic acid contains 3.5-5% nitrogen. After acid hydrolysis about 50-60% of the nitrogen goes into solution in the form of amides and mono-and diamino acids. The molecular structure of humic acids has not been determined. According to the available data, more than one humic acid exists. Dragunov (1948) found that two samples of humic acid, one obtained from peat and the other from chernozem, differed from each other in their chemical composition, in the amount and structural type of their functional groups, as well as in the structure of their nuclei, Bremmer (1955) subjected samples of humic acids, obtained by him from nine different soils, to chemical analyses. Each sample of the acid was analyzed for total nitrogen, ammonia-nitrogen, amino-nitrogen, and α- amino-acid nitrogen. The solutions obtained after hydrolysis were analyzed by paper chromatography for amino acids.

It was found that the samples of humic acids studied differed from each other in the composition of their nitrogen compounds and amino acids. Alkali extracts contain much of the nitrogen in the form of acid-soluble nitrogen compounds. About 20-60% of the nitrogen does not dissolve after acid hydrolysis. From 3-10% of the nitrogen is in the form of amino sugars. Nineteen amino acids were identified by means of paper chromatography: phenylalanine, leucine, threonine, isoleucine, valine, alanine, serine, aspartic acid, glutamic acid, lysine, arginine, histidine, proline, hydroxyproline, α- amino-butyric acid and others.

Humin and humic acids are decomposed by bacteria and fungi, especially by actinomycetes. Many actinomycetes grow well, bear fruit, and form antibiotic compounds on media containing humic acids as a sole source of carbon and nitrogen. Many forms of bacteria also grow on humic acid substrates.

Crenic acid was first found in spring water. According to Vil'yams, it is formed by fungi under aerobic conditions during the decomposition of forest vegetation and forest litter. Its properties differ sharply from other humic acids. It is colorless, highly soluble in water and acids, is not subject to denaturation and forms salts which can be crystallized.

Crenic acid possesses sharply pronounced acidic properties. According to Vil'yams, it can raise the soil acidity to such an extent that the activity and growth of many microorganisms is arrested.

It is difficult to accept this assumption. Organic acids as such are by themselves nutrients for many forms of microorganisms. It is quite clear, therefore, that their accumulation in the soil will be accompanied by an increase in the number of microbes.

Owing to its solubility, crenic acid penetrates deep layers of soil and there, combining with bases, forms crenates. They are harmless to microorganisms and are utilized by them as nutrients. Crenates are highly soluble in water, are easily leached from the soil, and may find their way either into ground or surface water. Thus, due to the high solubility of crenic acid and its salts, their accumulation in large concentrations is prevented.

Crenic acid may be reduced by nascent hydrogen with the formation of apocrenates. The reduction is carried out with the participation of anaerobic bacteria. Apocrenates are the salts of apocrenic acids. They have not been obtained in pure form. The salts of monobasic cations are highly soluble in water. Calcium apocrenate is slightly soluble in water and apocrenates of trivalent metals—iron, mangane se, and aluminum—are completely insoluble. These compounds are deposited in the soil in the form of voluminous amorphous sediments.

Crenic and apocrenic acids (fulvo acids) are widely distributed in soils. Their properties vary according to the soil. Kononova (1953) found that the acids from podsols differ from those of krasnozems.

The diversity of the natural conditions of soil formation both of a geographical and an ecological character, influence humus formation as a whole and, in particular, the composition of its individual components: humic, ulmic, and fulvo acids and other organic and organomineral compounds.

V. V. Dokuchaev was the first to point out the regular nature of the formation and transformation of humus under the varying conditions of different soils. climates. and zones. P. A. Kostychev and V. R. Vil'yams conceived the idea of the regularity of humus-compound formation in soils, in relation to the vegetative cover arid biochemical activity of the microflora.

Later investigations proceeding from chernozems to podsol soils, proved the regularity in the formation of the individual components of humus. Tyurin. (1949). developing the thesis of Dokuchaev on the basis of data from the literature and the results of his own investigations, showed that the geographical regularity of humus formation manifests itself not only quantitatively but also qualitatively.

As a rule, the humus of the coniferous forests of the northern and central belt of the USSR and, in general, of the podsol soils is of a bright color, it contains few stable humic and ulmic acids but many compounds highly soluble in water which are easily leached from the soil, e. g., crenic acid and apocrenates. Their concentration in podsol soils is 2-3 times higher than that of humic and ulmic acids (Kachinskii, 1956). In southern steppe regions having a grass vegetation, the soils contain humus with a different ratio of humic and fulvo acids.

The composition of humus in various soils also differs. Chernozem-type soils contain humic acids of different properties from those of podsol soils. Kononova (1956) showed the regularity in the variations of humic acids in the main soil types of the USSR. She found variations in the elementary composition of the acids, their optical density, and their distribution. The humic acids of chernozem soils are the most highly condensed, they are followed by the humic acids of the dark-gray forest soils, chestnut soils, and the bright-gray soils of the serozem; the humic acids of podsol soils and krasnozems are but weakly condensed, By applying the methods of X-ray structural analysis, the author determined the main structural outlines of humic and fulvo acids, which varied in relation to the type of the soil. These investigations disclosed the unity of the soil-forming process. While studying the genesis of humus and its components in various soils, Ponomareva (1956) reached analogous conclusions.

Investigators express three different points of view on the mechanism of humic-acid synthesis (Kononova, 1951). The majority of workers consider that the formation of these compounds is outside the activity of microorganisms. This was criticized by Kostychev and then by Vil'yams.

At present, microorganisms are considered to play an increasingly important role in the process of humus formation. Reistric et al., (1938,1941), by means of molds, detected the formation of compounds of the aromatic quinone series from sugars.

These investigations stimulated the study of products of microbial metabolism; products which could serve as building material for the synthesis of humic acid. At present, many foreign (especially German) and Soviet investigators are busy studying microorganisms, their metabolic products, and the synthesis of humus compounds.

Great attention has been drawn to molds, actinomycetes, and heterotrophic bacteria as producers of humus-like compounds. While studying the products of bacterial metabolism, Martin, J. (1945) found that about 30% of the humus is synthesized at the expense of bacterial polysaccharides of the uronic type. The most stable of them, "levan", is formed by sporogenous bacteria *Bac. mesentericus* and *Bac. subtilis*.

Flaig (1952) isolated 42 cultures of actinomycetes from the soil which, under certain conditions, form a dark-brown or almost black humus-like compound. Kuster (1950-1952) concentrated his attention on fungi which produce compounds similar in color and certain chemical properties to humin substances. Laatsch. Hoops, and Bieneck (1952) found that the fungus *Spicaria* and certain actinomycetes, when grown on artificial protein media, are capable of forming a compound closely related to humin. Scheffer and Twaditmann (1953), Plotho (1950), Laatsch and others succeeded in finding a medium in which, under given conditions, fungi or actinomycetes formed substances of the phenol type. These investigators assumed that the oxidation-reduction systems—quinones $<=>$ polyphenols—are in a state of continuous activity in the living cell, being oxidized by polyphenol oxidases and reduced by dehydrases. With the cessation of respiration the quinones are released from the cell, being irreversibly oxidized; they then combine with organic nitrogen compounds (protein decomposition products) to form humic acids. Consequently, according to the above-mentioned authors, the reaction of quinones with microbial nitrogen compounds is the basis of humus formation.

Wilts (1952) noticed that humic substances are formed from various organic compounds. The building blocks of the humus particles may be products of decomposition of lignin and of tannic compounds—aromatic compounds of the phenylpropans series, easily hydrolyzed carbohydrates (cellulose and others), and proteins which are subject to complex transformations as a result of bacterial metabolism.

The works of Soviet investigators Mishustin, Gel'tser, Rudakov, Kononova, and others should be mentioned. Mishustin (1938) studied the formation of humus substances upon self-heating of grain; Gel'tser (1940), upon the decomposition of fungi; Rudakov (1949) ascribes the main role in humus formation to pectin compounds. Troitskii (1943) assumes that humic acids are formed by microbes from decomposition products of vegetative residues. Tepper (1949, 1952) has shown that humin substances are formed at the expense of pigments formed by fungi and actinomycetes.

Kononova (1951), in her monograph, proposes that various plant residues and products of resynthesis, as well as the microbial protoplasm participating in the process of humus formation, may serve as sources of humus. According to her. the primary molecule of humic acid emerges as a result of the condensation of aromatic compounds with an amino acid or polypeptide. This process takes place with the participation of microorganisms under the conditions of biocatalysis maintained by the oxidative bacterial enzymes. As a result, nitrogen-containing compounds of a cyclic structure are formed.

Radioactive Substances of the Soil

Among the mineral elements of the soil a special place is occupied by radioactive substances: radium, uranium, thorium, and others. According to Baranov and Tseitlin (1941) their content (weight %) in different soils in as follows:

	Ra ($\times 10^{-11}$)	*U* ($\times 10^{-5}$)	*Th* ($\times 10^{-4}$)
Krasnozem, Batumi	6.71	20.13	9.19
Desert serozem	2.96	8.8	2.61
Bright-chestnut	8.22	24.66	5.63
Medium podsol loams, Moscow Oblast'	8.88	24.66	5.63
Dark forest	7.45	22.35	5.99
Podsol, Leningrad Oblast'	9.46	28.38	4.79
Loamy chernozem	9.08	29.24	5.14
Mountainous tundra, Khibiny	7.46	22.58	4.10
Marshy tundra, peat	1.94	58.3	9.5

The biological significance of natural radioactive elements remains unknown. It should be assumed that it is of considerable importance for the plant, animal, and microbial population of the soil. Existing data show that these substances in small concentrations activate biological processes, increase metabolism, and exert a positive influence on the growth of plants. The natural radioactive substances of soil find their way into plants, may concentrate there, and cause definite effects (Drobkov, 1951; Vlasyuk, 1955; Popov, 1956, and others). The biological action of radioactive substances (radium, uranium, radium emanations and others) has been studied for a long time by microbiologists. Nadson et al. (1920, 1932), Filippov (1932). and Rokhlina (1930, 1954) studied in detail some of the biological processes of yeasts, fungi, and bacteria caused by radium, radium emanations. X-rays, etc.

These authors were the first to establish the effect of radium and other sources of radiation energy in promoting genetic mutations. We (Krasil'nikov, 1938) have shown that various types of luminescent actinomycetes react differently to the radiation of radon. Certain species were more sensitive than others. Radon rays have a stimulating or suppressing effect on the growth of mycobacteria, actinomycetes, and proactinomycetes. According to our observations, pigmented cultures are more sensitive to radon than nonpigmented ones.

In recent years we have studied soil bacteria—*Azotobacter,* root-nodule, and some others and their relation to certain radioactive substances, such as radium, thorium, and uranium. It was found that the bacteria absorb these substances from the soil and accumulate them in their cells in considerable quantities which many times exceed the concentrations of these substances in the soil. Attention should be drawn to the fact that the degree of accumulation of radioactive substances in cells varies in different kinds of bacteria. Some kinds of bacteria, especially *Azotobacter,* accumulate radium in large quantities, others, in small quantities, or are completely devoid of this capacity. Even in the same genus different strains accumulate natural-radioactive substances to a varying extent.

The radioactive substances inside the bacterial cell stimulate growth and metabolism. Nitrogen fixation by *Azotobacter* is enlarged under the influence of radium and thorium. The ability of root-nodule bacteria to penetrate the roots of legumes and to form nodules is also increased (Krasil'nikov, Drobkov, Shirokov, and Shevyakova, 1955). As a rule, the activating doses of the substances studied by us cannot be detected by ordinary electronic counters (radiomer B-2 and others). The microorganisms are sensitive to irradiation by radioactive substances in doses which cannot be detected by modern instruments.

The microbial population of the soil as well as plants are adjusted to small concentrations of radioactive elements. High doses given to them artificially under experimental conditions are harmful. Minimal concentrations of radium, uranium or thorium which can be detected by electronic counters damage even the least sensitive species of bacteria. Under the influence of such doses the cells undergo degeneration, increase in size and deform, their protoplasm becomes coarsely-granular, vacuoles appear, and their reproduction slows down and eventually stops altogether. Similar changes were observed by Filippov, Shtern, Rokhlina, and other collaborators of Nadson, in yeasts, fungi, and certain plants when irradiated by X-rays, radon, or ultraviolet rays. The same picture of degeneration in yeasts, under the action of large doses of radium and other sources of radioactive irradiation, was noted by Meisel' (1955). His investigations led to the emergence of a scheme of consecutive damage to the structure and function of cells.

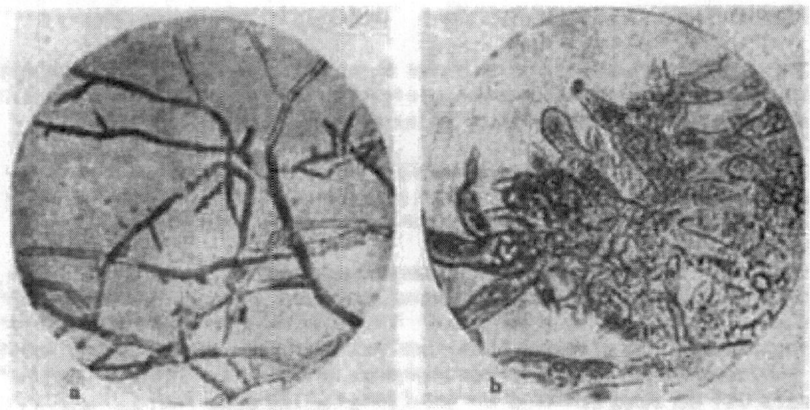

Figure: *The effect of radioactive compounds (U) applied in the minimum doses detectable by the electronic counter (B-2) on the culture of* Aspergillus niger:*a) control; growth on a medium without radioactive substances; b) growth on a medium containing uranium. Swollen hyphae of the mycelium with degenerative coarsely-granular protoplasm.*

As noted by Vernadskii (1926, 1929), radioactive substances possess free energy and continuously carry out considerable chemical activity in the soil. The energy of radioactive elements affects chemical and biochemical processes of microbes and organisms. Vernadskii stresses the fact that life in the biosphere originates from two energy sources: solar radiation and atomic radioactive energy. According to his calculations, only three radioactive elements, uranium, thorium, and radium supply the earth with heat, the quantity of which exceeds a thousand times that received by the earth's surface.

The biosphere of the earth accumulates dispersed radioactive elements and concentrates them on the surface, thus essentially changing the energetics of the whole population. It should be assumed that plants, animals, and microbes have, during their long evolution, acquired the ability to utilize these powerful energy sources. Analyses show that radioactive substances are present, to a larger or smaller extent, in all organisms and almost always in concentrations exceeding those in the surrounding environment. In many instances, plants contain ten times or hundreds of times higher concentrations of radioactive substances than the surrounding substrate (Vinogradov, 1932; Baranov and Tseitlin, 1941; Drobkov, 1951, and others).

The problem as to whether the organisms require the radioactive substances remains experimentally unsolved. Opinions are held according to which these substances, in small doses, do not play any role in the life of organisms and, in large doses, are harmful. Recently, data have accumulated which prove the reverse: small doses of radium, uranium or thorium stimulate the growth and increase the dry mass yield. Studies on the importance of natural-radioactive substances in soil fertility and in the life of plants and microbes are still inadequate. There are a number of observations which give reason to believe that these substances play an essential role in nitrogen fixation. A question arises as to the energy source needed to fix 100 to 150 kg and more of molecular nitrogen per one hectare of soil in one season. To fix such amounts of nitrogen and they are actually of this magnitude, *Azotobacter,* the most powerful nitrogen-fixing organism, requires 5-10 tons of glucose. Such quantities of this energy-yielding material are hardly to be found even in the most fertile soils.

Perhaps in this case the radioactive soil substances constitute the energy source which is indispensable for nitrogen fixation, as well an for many other processes taking place in the natural environment. The natural-radioactive substances deserve the most painstaking studies as biocatalysts on the earth's surface. When they enter into the chemical composition of living organisms, it should be assumed that they are not destroyers but creators, participating in many transformations and stimulating various enzymatic processes.

The Adsorption Capacity of Soils

Soil is known to adsorb various substances. There are a number of forms of soil adsorption: mechanical, physical, chemical, biological, and physicochemical.

Mechanical adsorption. The soil, as any other porous body, retains particles present in the filterable liquid. In other words the soil acts as a filter. Ordinarily, the size of retained particles exceeds the size of the soil pores but even smaller particles can be retained.

Physical adsorption is linked to the phenomenon of surface tension and is manifested by the fact that increase or decrease in the molecular concentration of compounds in the solution takes place an the surface of particles.

Physicochemical or exchange adsorption—consists in cation exchange. The cations from the solid phase of the soil are being exchanged for an equivalent quantity of cations present in the surrounding soil solution.

Chemical adsoerption expresses itself in adsorption of certain ions from the soil solution, which form insoluble salts in soil. Consequently, a precipitate is formed which enters into the solid phase.

Thus, for example, the ion of phosphoric acid precipitates in the presence of calcium salt (carbonic, hydrofluoric. or sulfuric acid). An insoluble salt of tricalcium phosphate is obtained. The latter precipitates and enters into the composition of the solid phase of the soil.

Biological adsorption according to Gedrolts is characterized by the adsorption of compounds from the soil solution, by microbial cells and green plants.

The adsorption of microbial cells by the soil also belongs here.

According to Gedroits, the physicochemical adsorption is of the greatest importance. In his opinion it is conditioned by the soil-adsorbing complex, consisting of chemical substances capable of exchange reactions. These substances may be organic and inorganic compounds or colloids undissolved in the soil solution. The latter represent the smallest particles, in size less than 1μ more often from 1-100m μ which do not precipitate in water and pass through fine filters. They are only visible in the ultramicroscope.

Organic compounds of the soil such as the humic acids, organomineral and inorganic compounds such as aluminum-silicates, iron hydroxide, argillaceous minerals and others may exist in the colloidal state. They represent a finely dispersed system, the particles of which possess high surface-reaction capacity for adsorbing substances present in solution. The soil colloids can be divided into hydrophiles and hydrophobes. The first adsorb water molecules and hydrated ions of the soil solution on their surface. The latter do not absorb molecules of the liquid phase of the solution.

The soil colloids adsorb cations. This adsorption is an exchange process, since with the adsorption of these cations, other cations are being released in equivalent quantity.

The sum of all adsorbed or exchanged cations which can be eliminated from the soil is a constant value for a given soil (Gedroits).

It varies only with the acquisition of new properties by the soil and with the change of its essential nature.

The sum of adsorbed bases comprises the volume capacity of adsorption. It in expressed in milliequivalents per 100g of soil. The adsorption capacity varies in different soils. It is smallest in podsol soils and largest in chernozems. The former is conditioned essentially by a mineral adsorbing complex and the latter by the organomineral part of the soil.

Soils saturated with bases (chernozem, etc), contain magnesium and calcium in the adsorbing complex. Saline soils. besides these two elements also contain sodium. There are soils which are not saturated with bases. To these belong the podsol soils which contain hydrogen.

The adsorption capacity is conditioned by the composition, properties and degree of dispersion of the soil. The greater the number of small particles in the soil, the higher the specific adsorption surface. Soils having a large percentage of highly dispersed organic humus compounds possess higher adsorption capacity than soils poor in organic compounds. The adsorption capacity of humus is 150-250 milliequivalents and humic acid, 300 milliequivalents per 100g.

Soils possess exchange capacity not only in regard to cations but also anions. The adsorption of anions takes place in the soil in the presence of iron and aluminum hydroxides.

The adsorption of microbial cells by soil particles is also of great importance. This phenomenon has been inadequately studied and much of the data requires experimental verification; certain data are contradictory. Nevertheless, the little data available are of great interest.

Adsorption of Bacteria by Soil

It has long been noted, in the laboratory practice, that bacterial cells are adsorbed by various materials in powder form. Kruger, (1889) demonstrated the adsorption of bacterial cells from their aqueous solutions by coke, clay, brickflour, magnesium oxide and other substances. Later Eisenberg (1918) showed that bacterial cells can be adsorbed by animal charcoal. Michaelis (1909) noticed that different bacterial genera are adsorbed to a varying degree.

Bacterial adsorption by soil particles was demonstrated by Chudiakov, N. N. (1926) and his collaborators Dianova, Voroshilova (1925), Karpinskaya (1925) and others. These investigators have shown that the soil adsorbs considerable quantities of bacterial cells. According to Dianova and Voroshilova (1925), between 252 and 4,350 million bacterial cells par hectare are adsorbed depending on the kind of soil and generic peculiarities of the bacteria.

Table: *The adsorption of* Bact. prodigiosum *by different soils (number of cells in millions, per 5 g of soil)*

	Cells introduced	***Total adsorbed***	***% adsorbed***
Podsol exp. fields of Agr. Acad. im Timiryazev			
	58,600	4,470	58.8
	5,860	4,988	85.9
	58.6	52	90.4
Chernozem of the Voronnezh Oblast'			
	32,800	28,000	87.5
	16,400	16,200	98.8
	3,280	3,230	98.5
	328	327	99.7

The loam soils of the experimental fields of the Agricultural Academy im. Timiryazev adsorb *Bac. mycoides* 95.5%; *Bac. ellenbachensis* 57.5%; *Bac. mesentericus,* 40.5%; *Ps. fluorescens liquefaciens* 79.7 %; *Staph. pyogenes,* 80%; *Bact. prodigiosum,*98%; *Bact.coli,* 10-20%.

Novogrudskii (1936c) studied the adsorption of podsol soils of the experimental fields of the Agricultural Academy im. Timiryazev, the soils of the Moscow Botanical Garden and chernozem of the Voronezh Oblast'. We have compiled the result of these studies in the tables. The data on bacterial adsorption and the adsorption of fungi and actinomycetes.

Table: *Adsorption of bacteria by different soils (in %)*

Soils	% ***Bac.Mycoides***	% ***Bac. Mesentericus***	% ***Bac. megatherium***	% ***Bac. chroococcum***	% ***Bac. fluorescens***	% ***Bac. denitrificans***	% ***Bac. leguminosarum***
The timiryazev Agricult. St. (podsol)	71	10	61	64	8	36	44
Botanical Garden	82	76	62	44	20	20	45
Voronnezh Oblast'	99	99	93	95	50	82	88

Table: *Adsorption of spores of fungi and actinomycetes by different soils (in %)*

Soils	% *Asperg. niger*	% *Penic. glaucum*	% *Mucor mucedo*	% *Fus. sp.*	% *Botrytis cinerea*	% *Act. 154*	% *Act 105*	% *Act. 110*
Podsol	14	43	57	99	93	8	10	13
Botanical Garden	20	43	27	97	97	15	31	28
Voronezh chernozem	97	94	97	99	99	99	75	94

According to our data the Moldavian chernozem (medium loams, carbonate) adsorbs two to three times more cells of *Azotobacter* than the upper layer of the podsol soil which was previously under forest (Experimental Station of the Moscow State University, Chashnikovo, Moscow Oblast'). In the former soil, 4,000 million *Az. chroococcum* and 600 million *Az. vinelandii* were adsorbed per gram and in the latter soils 2,400 million and 200 million per gram respectively.

The adsorption capacity of soil varies according to the depth. The upper layers of the soil are characterized by higher adsorption capacity than the lower, The poorly cultivated soil of the Experimental Station of Chasnikov adsorbed *Az. chroococcum,* according to the different horizons, as follows:

a) Layer A-A_1; 80 % in May and 92 % in August
b) Layer A2 (10-20 cm); 50% in May and 53% in August
c) Layer B_1 (30-40 cm); 35 % in May and 25% in August
d) Layer B_2 (50-70 cm); 60% in May and 75% in August

Under the same conditions soil of the same type. but well cultivated, adsorbed cells of *Azotobacter* as follows:

- In the layer 0- 10 cm; 85 % in May and 93 % in August
- In the layer 10-20 cm; 80% in May and 83% in August
- In the layer 30-40 cm; 65% in May and 65% in August
- In the layer 50-70 cm; 78% in May and 87% in August

The adsorption capacity of soils is closely connected with their mechanical composition. Sand contains particles 1.0 to 0.25 mm in diameter and sand dust with particles 0.25-0.05 mm in diameter, adsorb bacterial cells weakly. Dust containing particles 0.05-0.01 mm 0.01-0.005 mm and 0.005-0.0015 mm in diameter adsorbs microbial cells most actively. The slimes of rivers and lakes having particles of 0.0015 mm and less in diameter are devoid of adsorbing capacity. since the size of the particles of slimes (1-1. 5 μ and less) does not exceed that of the ordinary bacterial cell. In such an environment bacterial cells are themselves adsorbants. The greater the adsorption of bacteria, the fewer the cells found afterward in the suspension.

The character and degree of adsorption of microbial cells by the soil is conditioned to a large extent by the qualitative features of the organisms proper, and their generic properties.

The degree of adsorption depends on their metabolic state and their vital potential. Some bacterial genera are adsorbed more vigorously and in a larger quantity than others. According to some authors, many nonsporiferous bacteria are adsorbed considerably weaker, by the same adsorbent, than the sporiferous bacteria or micrococci.

For example, podsol soil adsorbs bacterial cells as follows:

Bac. mycoides 71%

Bac. megatherium 61%

Az. chroococcum 64%

Ps. fluorescens 18%

Bact. coli 10%

Bact. denitrificans 36%

Rhizob. leguminosarum 44%

Gram-positive bacteria are adsorbed by the soil in larger quantitites than the gram -negative bacteria. Bogopol'skii (1933) gives the following data. Peat of medium decomposition adsorbs 74% of *Bac. mycoides,* 81% of *Urobact. pasteurianum,* 21-22% of *Bact. coli* and *Ps. fluorescens.*

The same results were obtained by Eisenberg (1918) while studying the adsorption of bacteria by charcoal and other adsorbants. According to his data, the adsorption of gram-positive bacteria, *Micr. pyogenes, Micr. candicans Sarcina lutea* and others was 500 times larger than that of the gram-negative bacteria, *Bact. coli, Bact. typhi, Ps. pyocyanea, Vibrio cholera* and others.

The degree of adsorption of microbial cells by the same soils depends upon the pH of the suspension from which the cells are being adsorbed. The spores of *Bac. mycoides* were maximally adsorbed at PH 4.5, with the increase of pH to 5.8-6.7 the percentage of spore adsorption decreases.

When the pH of the medium is raised to the neutral (pH 7.0), and higher into the alkali zone (pH 7.8), the degree of bacterial adsorption remains on the same level or even decreases slightly.

Table: *Bacterial adsorption by podsol soil at different pH(number of cells grown in the Petri dish) (according to Eisenberg, 1918)*

	pH of the suspension	***suspension***	***soil***	***adsorbed %***
Bac. mycoides				
	4.6	82	8	90
	5.8	66	33	50
	6.7	67	59	12
	7.4	63	46.5	27
	7.8	66	45.5	28
Bact. coli				
	4.4	384	311	19
	5.6	361	384	0
	6.3	384	178	35
	6.9	360	199	45
	7.5	358	294	20

The adsorption of *Bact. coli* under the same conditions is different. The highest adsorption takes place at a pH of 6.3-7.5; in an acid or strongly alkaline medium the cells are adsorbed much more weakly.

The adsorption capacity of the soil varies in relation to its moisture. Very wet soil adsorbs less bacteria. Upon rinsing with water a considerable quantity of adsorbed bacterial cells is desorbed and washed out, A single washing of podsol soils, according to our experiments, releases about 11% of *Az. chroococcum.* The larger the volume of water, and the more prolonged the elution of the soil, the more bacteria are washed out. Of the 2,900 million adsorbed cells of *Azotobacter* the following quantities were eluted:

- First elute, after one minute, 330 million (11.4%)
- Second elute, after two minutes, 54.0 million (0.18%)
- Third elute, after two minutes, 35.2 million (0.12%)
- Fourth elute, after two minutes, 0.2 million (0.007%)

The eluted bacteria comprised 14% of the total. One hundred ml of water were used for each washing per 5g of soil.

The elution of the bacteria is apparently, limited. Above this limit the bacteria are not desorbed even after prolonged washing. The quantity of desorbed cells in different soils varies. The desorption of bacteria by water from their natural surroundings is observed after rain or irrigation. This is closely connected with the seasons.

The adsorption capacity of soils varies during the vegetation cycle. According to the data of Novogrudskii (1937), less bacteria are being adsorbed in spring and autumn than in summer,

In April and September, in the podsol soils of the fields of the Agricultural Academy im. Timiryazev, 40-66% of bacteria were adsorbed and in the summer 60-90%. The seasonal variations of the adsorption capacity of the soil are conditioned not only by moisture but also by temperature. The less moisture in the soil, and the higher its temperature, the stronger the adsorbing capacity for microbial cells.

Samples of soils at 25% moisture per dry weight maintained at 0° and at 25°, adsorbed 57% and 68% *Bac. mycoides* respectively.

The adsorption of bacterial cells by the soil is a reversible process. Upon change of pH, temperature, moisture and other factors. the bacteria are desorbed.

An exchange adsorption is observed, similar to that of mineral substances. if the soil is saturated with one type of bacteria and is then saturated with cells of another more adsorbable kind, then an exchange of bacterial cells will take place. The formerly adsorbed cells will be released or displaced and will reappear in the suspension.

When two or more kinds of bacteria are simultaneously adsorbed, the more adsorbable ones will be preferentially adsorbed (Novogrudskii, 1936).

According to Chudyakov and collaborators, the adsorbed cells preserve their viability but their metabolism slows down or stops altogether.

Dianova and Voroshilova (1925) determined the biological activity of bacteria in strongly adsorbing soils and in sand. The substrates were wetted with nutrient mediums (such as peptone, glucose and others), were sterilized and inoculated with *Bac. mycoides, Bact. prodigiosum* and *Sarcina flava.* Their biological activity was determined by the CO_2 released.

Under all experimental conditions the authors noticed a strongly diminished liberation of CO_2 from the soil.

For example, in experiments with *Bac. mycoides* the following amounts of CO_2 were released: in sand, 106.3-126.8 mg, in soil, 0-7.8 mg; with *Bact. prodigiosum* 24-52 mg of CO_2 was released in sand and 0 in soil; *Bact. megatherium* released 66 mg of CO_2 in sand and 2.6 mg in soil; *Bact. coli* released 41-69 mg of CO_2 in sand and 10-23 mg in soil.

The stronger the adsorption of bacteria, the less their activity. The activity of *Bact. coli* in sand is 2-4 times higher than in soil, while that of *Bact. mycoides* and *Sarcina flava* is 10-30 times higher than in soil.

The activity of adsorbed bacteria increases with the rise of soil moisture. At 60% of total moisture capacity, 5.2 mg of CO_2 was released from the soil, at 100% of moisture capacity, 28.6 mg (experiments with *Bac. mesentericus).*

According to Lipman (1912), the nitrification process by bacteria in a clay soil proceeds slower than in sand. *Bact. proteus* releases 40% more ammonia nitrogen in sand than in clay; *Sarcina lutea* 80% and *Bac. mycoides* 87%.

According to our observations, cells in the adsorbed state reproduce quite actively. Thus, for example, after careful washing of the soil *Azotobacter* remained in the adsorbed state at a level of 55 millions per gram of a total 100 million per gram introduced. We washed the soil samples daily for a month. About 300 million cells per gram were washed out. However, after the last elution, a considerable quantity of cells remained in the adsorbed state. Thus, they increased daily by 10 million bacterial cells for each gram of soil.

Apparently, the process of adsorption of microbial cells by soil particles is also of a biological, and not on ly a physiochemical character, and it must not be considered only from the point of view of physical or chemical forces. Krishnamurti and Soman (1951), analyzing the literature data and their own investigations, reached the conclusion that the phenomenon of adsorption of bacterial cells is of a specific character. The percentage cell adsorption is conditioned by the adsorbent properties as well as by the generic properties of the microbe. The adsorption coefficient is strictly constant under given conditions. The authors even recommend the differentiation of bacterial species on this basis.

Adsorption of Products of Microbial Metabolism by Soil

There are almost no data in the literature on the adsorption by the soil of products of microbial metabolism although this problem is of considerable theoretical and practical interest. Microbes, as was pointed out above, grow abundantly in soils, proliferate, display high biological activity and synthesize and release various metabolic products into the milieu. Among these products there are many biologically active substances: enzymes, vitamins, auxins, amino acids and other biotic substances. Antibiotic metabolites, toxins, etc can also be found, Once these substances are excreted from the cell into the soil, part of them undergo decomposition and inactivation, another part is adsorbed by soil elements. The degree of adsorption of such active metabolites is unknown. It should be noted that the literature contains very little data on the adsorption of organic compounds by the soil. The adsorption capacity of the soils, as was stressed above, was studied mainly in reference to mineral elements: cations and anions. No attention was paid to organic compounds. However, these processes of interaction between the soil and organic compounds should provide an explanation for the formation of organomineral compounds which determine the essence of soil fertility or the formation of soils as such.

The available data on adsorption of organic compounds by the soil mainly refer to the problem of humus formation.

Kravkov (1937) introduced aqueous extracts of grasses and straw into the soil and observed their fixation. According to his data, the water-soluble plant compounds are adsorbed by soil particles to a varying extent which depends on the type and properties of the soil. A different adsorption capacity was recorded for each soil. The organomineral compounds so formed are considered by the author to be the humus of the soil.

Persin (1944) introduced aqueous extracts of fresh straw and hay of various grasses as well as extracts of straw and hay, after they had been subjected to decomposition by microorganisms. It was found that the water-soluble extracts of fresh straw and hay are not adsorbed by the soil and that extracts of decomposed straw and hay are adsorbed to a varying degree, depending on the stage of decomposition. The greatest adsorption of water-soluble compost substances was observed after 75 days of decomposition at the optimal temperature for microbial activity.

According to the observations of the author, chernozems adsorb organic compounds in greater quantities than podsol soils. The adsorption capacity of soils for organic substances is conditioned by their mechanical composition. The larger the clay fraction in the soil, the greater its adsorption capacity, and, consequently, it retains the adsorbed substances more tenaciously.

It should be noted that Kravkov, Persin and some other authors (Chizhevskii and Makarov, 1939) carried out their experiments in nonsterile soil. Consequently, a great part of the introduced organic compounds (if not all of them) was decomposed by microorganisms and was lost to the investigators. The real magnitude of adsorption in these experiments cannot be precisely determined.

It should be noted that in the investigations of Persin, the soil adsorbed only those water-soluble organic substances which are obtained from decomposed plant residues, i. e., substances formed by bacterial activity.

Simakov (1938) carried out experiments with tannin and xylan. These substances were differentially adsorbed by the soil, xylan less than tannin. The author in his work during 1944 carried out experiments on adsorption of amino acids and sugars by lignin, which represented one of the components of the soil complex. The experments have shown that the afore-mentioned substances were strongly adsorbed by lignin and equally strongly retained. During this process their properties changed; they became more stable.

According to Simakov (1944), the amino acids asparagine and glycine adsorbed by lignin are decomposed slowly by microorganisms.

A considerable number of papers have been devoted to the adsorption of humic substances by the soil (Zyrin, 1945; Khan, 1950-51; Aleksandrova, 1944; Tyurin, Gutkina, 1940, and others). These investigations have shown that humic substances form stable organomineral compounds with the mineral parts of the soil, the bond between the mineral particles and organic substances of the humus may be of a physical or a chemical nature.

We (Krasil'nikov, 1954c) have tested antibiotics of actinomycetes, bacteria and fungi.

Antibiotics, due to their specific antibacterial action are easily detectable, and can be found in various natural substrates, as for example in the soil. They are, therefore, convenient objects for the determination of the adsorption capacity of soil particles.

Antibiotics were introduced into various soils and their adsorption determined. We tested penicillin, streptomycin, globisporin, aureomycin, terramycin, subtilin, gramicidin, and other antibiotics, and have shown that they are adsorbed in considerable quantities. For example, we introduced streptomycin at a concentration of 2,000 units/g; after some time 1,120 units/g were adsorbed by the chernozem, 1,800 units/g by podsol, 1,080 units/g by the serozern and 1,540 units/g by krasnozem.

Similar quantities of globisporin were adsorbed by the different soils. Penicillin was adsorbed as follows: 380 units/g by chernozem, 280 units/g by podsol soils, 380 units/g by serozem, and 200 units/g by krasnozem.

Aureomycin and terramycin as well as antibiotics of bacterial origin such as subtilin and gramicidin were adsorbed by the afore-mentioned soils in varying quantities. The antibiotic 1609 was only adsorbed by the podsol soils, and then only in a very small quantity, 20-30 units/g. This antibiotic was not held by any other soil.

Thus, the various soils adsorb different quantities of antibiotics, however, the nature of their adsorption is different from that of mineral compounds.

Soils poor in humus (podsols, krasnozems) adsorbed antibiotics in considerably larger quantities than soils rich in humus. Different layers of the same soil possess different adsorption capacities.

We have studied the streptomycin adsorption capacity of the podsol soils (Experimental Station Chashnikovo, Moscow Oblast') of cultivated and noncultivated soils.

Table: *Adsorption of streptomycin by the podsol soils at various layers (units/ g)*

Soil	***A_0 Layer***	***A_2 layer***	***B_1-B_2 layers***
Forest mixed	1,300	700	5,400
Meadow	1,800	1,200	7,400
Roima* of the river Klyaz'ma	3,000	2,400	3,100
Glade	1,500	300	3,000

*[Russian term designating a very low and broad river terace,which may be flooded at the highest water mark. Flood basin but very extensive.]

The B_1-B_2 layer has the strongest adsorption capacity and the layer A_0-A_2 the smallest. The degree of adsorption of antibiotics by the soil does not depend solely on the soil properties but also upon the properties of the antibiotics themselves. In one and the same soil, for example in chernozem, we have observed the following adsorption:

	Units/g	μ g
Streptomycin	1,120	2.2
Globiosporin	1,080	1.8
Terramycin	900	1.0
Pennicillin	380	0.1
Preparation 1609	0	—

The antibiotics in the adsorbed state preserve their antibacterial activity for some time. The period for which a given antibiotic preserves its activity depends on the soil and the properties of the antibiotic itself.

For example, in same soils penicillin remains active for 20-30 hours, in others 2-3 hours; terramycin remains active for 3-5 days in podsol and 1-2 days in chernozem. Some antibiotics (preparation 1600) lose their activity immediately.

Organic compounds adsorbed by the soil undergo various transformations; they are decomposed, inactivated and disappear, They are replaced by other compounds.

The adsorbed fraction of the antibiotics retains its antibacterial properties for a more prolonged period than the antibiotics in the free state present in the soil solution. For example, free streptomycin disappears from the podsol soil after 10-12 hours, while adsorbed streptomycin is preserved for more than 30 hours. In the serozem, the nonadsorbed streptomycin is inactivated after 20-25 hours; its absorbed fraction can be detected even after two days. A still greater difference was observed in the experiment with aureomycin. In podsol it can be detected in the free state after 20 hours, in the adsorbed state after 5 days In the serozem the free antibiotic is preserved for no more than two days, while in the adsorbed state it retains its activity for more than seven days.

The antibiotics in the soil are partially inactivated by the soil solution and by microorganisms.

Not only antibiotics but also other microbial metabolites, as well as intermediate decomposition products of plant residues, and various compounds of humus, are adsorbed by the soil.

The biologically active metabolites present on the surface of soil particles exert a great influence on their physicochemical state. The soil particles holding the substance on their surface gain new properties.

The presence of living microbial cells adsorbed on the soil particles should be regarded as a complex system of biotic-mineral complex. Each soil particle carries elements of living organisms, the study of which is of the utmost importance to soil biologists.

The Microflora of the Soil

Microorganisms are an integral part of the soil. If the soil should lose these organisms it would lose its main property—fertility—and it would turn into a dead, barren, geological body.

The soils are inhabited by numerous representatives of the microflora—bacteria, actinomycetes, yeasts, fungi, algae, protozoa, insects, worms, and others. Besides, there are in the soils various ultramicroscopic organisms: bacteriophages and actinophages.

No accurate data on the numbers of microorganisms in the soil are available. Methods for the detection of the entire soil population are not available. The existing methods give only a relative idea of the density of the microbial population.

Two essentially different methods are employed for the quantitative estimation of microbes in the soil: a) determination by means of soil inoculation of artificial media-liquid and solid, b) direct count of cells. These two methods give different data on the quantitative aspect of the microbial population of the soil.

In practice, the inoculation method is more extensively used. There are various methods of inoculation and media.

The amounts of micoorganisms detectable in the soil vary, depending on whether they are inoculated into a solid or liquid medium, or inoculated by sowing the surface of agar medium, or dispersed in aqueous suspension by serial dilutions. Inoculation with soil is often carried out by placing small soil particles on agar medium.

In all instances the number of bacteria grown on agar media is smaller than upon growth in liquid media inoculated by the method of serial dilutions.

Data from the literature on the amount of bacteria, actinomycetes and fungi in the soil are obtained, in the majority of cases, from growth on agar media. According to these data, the number of bacteria per gram fluctuates within the range of from several tens or hundred thousands to many millions depending upon the soil composition and the medium (Starkey, 1929, 1931, 1955; Gray and Thornton, 1928; Clark, 1940; Timonin, 1940-41, Waksman, 1952; Jensen, 1934-36; Mishustin, 1956, and others).

Thom (1938), summing up data from the literature and the results of his own investigations on the quantitative determination of the bacterial population of the soil, considers that the total number of bacteria in one gram of soil reaches 50 millions. Since the greatest number of bacteria is concentrated in the plant rhizosphere many authors give data on microbial composition of this zone. Starkey showed, by the method of counting on agar media, the presence of 199 to 3,470 million bacterial cells per gram, depending upon the species of plant.

Humfeld and Smith (1932), counted 5-8 billion bacteria in one gram of soil with green fertilizer. Clark (1949) found 5 billion bacteria per gram of well-fertilized soil and also in soils under a mixture of grasses. Rippel (1939), analyzing the soils of Germany, and Feher (1933) analyzing the soils of Hungary and Austria, counted from one hundred thousand to 5 billion bacteria per gram of soil, depending upon soil composition and climatic conditions.

Nonfertilized soils have a smaller microbial population; ranging between hundred thousands and millions, but on the average 3-7 million per gram. Bunt and Rovina (1955) counted from 400,000 to 15 million bacteria per gram in the subartic soils of Iceland.

We have obtained similar figures. We have counted from several hundred thousands up to 15 million bacteria per gram, in the soils of the Kola Peninsula, the Islands of the Arctic Ocean and in mountainous soils of Pamir and Caucasus. The podsols of noncultivated or poorly cultivated soils contain, according to our investigations, 300 thousand to 10 million cells per gram of soil. Chernozems rich in humus contain 10-1,000 million cells per gram. Similar data were obtained by many other investigators studying various soils.

Higher counts (ten and hundred times higher) are obtained by the method of inocculation into liquid media and by the serial dilution method. For example, soils poor in organic compounds (podsols) gave from one to 100 million cells per gram, and fertile soils (chernozem) from one hundred to 1, 000 million bacteria per gram estimated by the solid-media method, meat-peptone agar (MPA). The method of inoculations on liquid media, meat-pe ptone broth (MPB) revealed to 10-500 million and 1,000-10,000 million bacteria per gram of soil respectively.

While studying the rhizosphere soil of lucerne in Central Asia (serozems) we have found (by the method of serial dilutions 50-100 billion bacteria per gram, and Raznitsina (1947) and Korenyako (1942) obtained even higher numbers. Such high figures are constantly obtained during the investigation of plant rhizosphere under given conditions.

Such high indices of microbial population throws doubt on the accuracy of the methods employed. Experiments especially designed to check this method were carried out. We have assumed that the high figures obtained by this method can be explained by the adsorption of bacterial cells on the walls of pipettes and with their subsequent elution (desorption). Experiments have shown that adsorpton of cells does indeed take place. The number of bacteria is 2-5 times, and sometimes even 10 times less if the pipettes are changed upon each dilution. The numbers obtained, when the pipettes are changed. are of an order of 1-10 billion per gram of soil, upon dilution of the soil with one pipette the numbers increase to 5-100 billion per gram of soil.

We checked the trustworthiness of this method by using pure cultures of *Bact. prodigiosum, Ps. fluorescens, Mycob. rubrum, Az. vinelandii* and *Bac. subtilis.* Aqueous suspensions of these bacteria were diluted with and without change of pipettes.

The number of bacteria in billions per 1 m/ obtained in such an experiment are as follows:

	With change of pipettes	***Without change***
Bact. prodigiosum	35.5	38.1
Ps. flourescens	22.8	100
Mycob. rubrum	1.0	4.5
Az. vinelandii	2.1	2.5
Bac. subtilis	3.9	7.3

In this experiment the change of pipettes lowered the number of bacteria 1.5-4 times depending upon the bacterial species. Similar lowering of bacterial numbers was observed on studying soil samples. The difference in the numbers is more pronounced if the number of bacteria in the soil is large. Hundred billions and more of bacteria were detected in the rhizosphere of lucerne grown in the Vakhsh valley when the soil suspension was diluted with one pipette; the number was five times less. when the pipettes were changed. The control soil contained 100-500 millions per gram upon dilution with one pipette, 50-150 millions were counted upon dilution, when the pipettes were changed, i.e., 30-50% less.

According to Vinogradskii, the method of direct counting of the soil bacteria also gives higher numbers than the method of growth on agar media, i.e., approximately the same are obtained upon serial dilution, or even higher.

Table: *The number of bacteria detected by various methods in the plow layer of soils under perennial grasses (in thousands/ g)*

Soil	***Direct count method***	***Liquid inoculation***	***Solid inoculation***
Podsol soil, field	560,000	500,000	7,500
Turf-podsol soil, garden, Moscow Oblast'	6,800,000	5,600,000	15,600
Chernozem, Moldavia	8,700,000	7,200,000	25,000
Chestnut soil, Trans-Volga region	3,500,000	1,000,000	9,500
Serozem, Central Asia	9,300,000	7,500,000	90,000

The difficulty of the method of direct counting is that the smears contain living cells and dead particles of the same soil. and they cannot be differentiated with certainty. The soil always contains a large amount of small particles which can be stained and thus become indistinguishable from the bacteria themselves. It is especially difficult to tell the coccoid cells from the small globular bodies and granules.

In recent years some investigators attempted to use fluorescent dyes for the differentiation of bacteria from the dead soil particles. Burrichter (1953). employed acridine-orange for the staining of soil smears and studied them under a fluorescent microscope in ultraviolet light. In soils rich in humus (9. 98 %)the author counted 9,453 million bacteria per gram of soil and the total number of microbes was 18,331 million bacteria per gram of soil; in soil poor in humus (1.80%) 1,230 million bacteria per gram were found. The number of colonies of slimy bacteria in the first soil amount ed to 157 million per gram. Soil, fertilized with compost. contained a total number of microbial cells of about 16,132 million per gram, and soils poor in organic substances 25 million per gram. Strugger (1948, 1949) found from 1,038 to 8,640 million bacterial cells per gram of soil employing the same method of fluorochrome staining.

It should be noted that the method of fluorescent staining also has shortcom -ings. The green color of living cells or the red color of dead cells and other particles

may often depend not on the cell viability but upon many other factors, such as the concentration of the dye, pH of the medium, temperature and others (Krasil'nikov and Bekhtereva, 1956). It is sometimes difficult to say what are the green or especially the red-stained bodies; are they living bacterial cells or dead, or soil particles.

The method of direct microscopy of soil smears (Kubiena, 1932) is also of little use for the quantitative estimation of cells. As can be seen from the given data, the existing methods of microscopic analysis are inadequate for quantitative determination of microbial numbers in the soil and for the determination of their forms. Therefore, when comparing data obtained by employing one of the aforementioned methods, the investigators limit themselves to relative figures.

The bacterial numbers vary in different soils, according to their fertility and nutritional qualities. The more fertile the soil, the richer it is in humus, the denser its microbial population. The podsol soils (Moscow Oblast) contain, in well-cultivated fields, 3-10 millions per gram, and the chernozem soil of the Kuban, contains (similar method of counting) 15-50 million bacterial cells per gram of soil.

One and the same type of soil also varies in the amount of microbes it contains. The podsol soils, not well cultivated and poor in humus, contain 500,000 to 1.5 million cells per gram and in some cases only a few thousand per gram (the soils of the Kola Peninsula). Well-cultivated, systematically fertilized soils contain 3-25 million cells per gram. Garden soils, as a rule, are richer in microbes than the soil of fields.

Virgin soils contain less microbes than cultivated soils.

Table: *The number of bacteria and actinomycetes in various soils (in thousands / g on meat-peptone agar, in Petri dishes)*

Soils	*Bacteria*	*Actinomycetes*
Podsol containing iron, Kola Peninsula	10-30	5-25
Podsol of Moscow Oblast' from under forest	100-300	70-100
Podsol of Moscow Oblast', garden	1,000-10,000	500-1,000
Chernozem, Kuban', under wheat	5,000-15,000	400-800
Serozem, Central Asia, virgin soil	850-1,500	600-1,000
Serozem, Central Asia, under lucerne	3,000-10,000	500-800
Chestnut soils, Trans-Volga region, virgin soil	400-1,500	450-860
Chestnut soils, Trans Volga region, under lucerne	5,000-15,000	500-1,000

The upper layer of the soil s richer in microbes than the deeper layers. For example, we have found the following amount of bacterial cells in podsol soils of the experimental fields of the Academy of Agriculture im. Timiryazev:

- in the layer 0-20 cm deep, 5.7 million/g
- in the layer 20-35 cm deep, 2.4 million/g
- in the layer 40-60 cm deep, 0.5 million/g
- in the layer 80-100 cm deep, 0.001 million/g

In the root zone of the vegetating plants, in other words, in the rhizosphere, the soil is saturated with bacteria to a greater extent than in the zone outside the roots. The vegetative cover, as will be shown later, exerts a strong influence on the concentration of microbes in the soil. The number of microorganisms in the soil varies with the season. According to the literature and our own data, their total number in winter is smaller than in summer. This is especially noticeable in the soils of the north.

The analysis of soils of Severnaya Zemlya and other islands of the Northern Ocean showed that in May, when the soil was still in the frozen state, it contained tens of thousands of organisms per gram and in August many millions of bacteria per gram.

Table: *Seasonal variations of microbial numbers in the soil of the Severnaya Zemlya (in thousands/g, counted on meat -peptone agar MPA)*

Soil sample	***May***	***August***
Sector I, loam	23	1,340
Sector II, loam	40	4,380
Sector III, loose calcareous soil	91	16,600
Sector IV, loam	14	3,600
Sector V, loose calcareous soil	112	6,600

The number of microbes in the soil of the temperate zone is greatest in spring, smaller in summer, it increases somewhat in autumn.

The data on numbers of bacteria in winter are few and contradictory. The majority of investigators think that life in the soil stops altogether during the winter. A considerable number of microbes die from cold and their total number decreases.

According to our observations, the microbial activity does not always cease in winter. Under a deep snow cover the earth is not always frozen and in such a soil microbiological processes take place. This can be found by studying the growth dynamics of individual species of actinomycetes. Korenyako has shown that during the winter months of 1952-1954 certain species of actinomycetes *(A. globisporus)* grew more abundantly, in Moscow Oblast' soils, than during the summer and autumn. Besides, certain biochemical processes, leading to detoxification of the soil take place in winter (Krasil'nikov, Korenyako and Mirchink, 1955).

The vigorous growth of microbes in spring is, according to our opinion, not only caused by the warm temperature and by moisture, but also by other factors, First, the toxins are inactivated or decomposed in winter due to low temperature. Second, low temperatures, as was noted above, stimulate the growth and activity of microbes. in addition, many soil nutrients under the action of low temperature, change and become more available to microbes. It was pointed out above that microbial growth in the soil depends on the presence of organic substances of humus. This is not always true. The amount of organic substances in the soil may be very high (peats, marshy soils) while the growth of microorganisms is rare. Not infrequently a reverse picture may be observed. Certain primitive soils of mountainous regions are poor in organic substances and at the same time rich in bacteria. The concentration of microorganisms in the soil depends mainly on the presence of such organic substances as can be easily utilized by bacteria. There are fresh plant and animal residues and products of their primary decomposition which have not yet been transformed into humus, as well as a number of products of synthesis, etc.

Of great importance for the life of microbes are organic growth factors; vitamins, auxins, various biotic elements and substances which suppress their growth and multiplication. Small doses of these substances markedly enhance the growth and multiplication of microbial cells as well as that of plants, by promoting various biochemical and physiological processes. This part of the organic compounds, or soil humus, is, in our opinion, of the greatest importance, and a correlation should be found between their quantity and the total number of the microbial population. Unfortunately, such a correlation is very difficult to study and has not as yet been methodically worked out.

Adsorption should be taken into account in the determination of microbial numbers in the soil. The data of observations and experiments given above showed the degree of bacterial adsorption by soil particles. Bacteria in the adsorbed state can be found in tens, hundreds, millions and billions in one gram of soil. The method employed by us (inoculation on media) in most cases accounts only for microbes in the free state as well as for a fraction of those adsorbed. The majority of adsorbed cells remains unaccounted for; the number differing from case to case.

The majority of investigators give data obtained by analysis of dry soil samples. Naturally, these data are far from the real figures. It is known that the number of microbes is decreasing in dry soil. During prolonged storage a large number of microbial cells die. Sometimes, upon drying, the total number of bacteria decrease by a factor of 2-3 and not infrequently 5-10 times.

Table: *The decrease in bacterial cells during dry storage in the laboratory (in thousands per gram)*

Soils	***Fresh samples***	***Samples after 10 days storage***
Chernozem of the Kuibyshev Oblast'	500,000	50,000
Serozem of the Uzbek SSR	150,000	45,000
Podsol of the Moscow Oblast'	3,500	1,500
Chestnut, Trans-Volga region	60,000	10,000
Severnaya Zemlya*	9,300	1,300

* Samples of the soil of Severnaya Zemlya (taken in August) were analyzed the same day and then after one month.

Upon storage of samples in the dry state the qualitative composition of the bacteria also changes. Some bacterial genera disappear almost completely, others remain in small quantities, still others do not decrease in numbers at all.

Actinomycetes and then mycobacteria are the most stable in this respect. The highest percentage of destruction is noted among the bacteria.

Table: *The survival of various groups of microbes upon storage in dry soil (in thousands/g)*

Soils	*Sporiferous bacteria*	*Nonsporiferous bacteria*	*Mycobacteria*	*Actinomycetes*
Chestnuts of the Trans-Volga region				
Fresh	1,500	56,000	1,000	1,500
Dry	450	5,000	900	1,000
Podsol of Moscow Oblast'				
Fresh	650	5,500	850	1,250
Dry	325	1,400	600	980
Chernozem of Kuibyshev Oblast'				
Fresh	2,500	400,000	25,000	25,000
Dry	280	46,000	16,000	26,000

In those cases where the soil dries up slowly, an increase in number of actinomycetes and certain species of mycobacteria is observed. These organisms can grow in soil of minimal moisture, when the growth of other microorganisms ceases (Krasil'nikov, 1940c).

Differences in survival capacity have been observed not only in different groups but also in different species, and even different strains of the same microbe show different ability to survive. According to our observation, cultures of *Bact. herbicola, Az. vinelandii,* nodule bacteria of soya and *Ps. aurantiaca,* die out rapidly in the dry soils of podsol (Moscow Oblast') and in virgin serozern soils. Of some hundred million cells only a few (10-100 cells/g) remained viable after two weeks storage. Mycobacteria such as *Mycob. rubrum* and some other species remain viable in considerable quantities (100,000 cells/g and more.)

Not all strains of *Azotobacter,* in dry samples of soil, die out at the same rate. Of twenty cultures of *Az. chroococcum* studied, eight strains of *Azotobacter* survived in considerable numbers—up to 10% and more of the cells. Of 10 strains of root nodule bacteria of lucerne only about 10% of 4 strains survived in dry samples of serozem soils; in the podsol soil only one strain survived and then only in a negligible amount (0.5% and less). The sporiferous bacteria show the same diversity as far as their survival capacity is concerned. About 80-90% of *Bac. megatherium* dies out in dry podsol soils of the Moscow Oblast', and 30-40% of *Bac. subtilis* and *Bac. mesentericus.* Only 5 strains out of 20 of the latter, when isolated from various podsol soils, were resistant to storage in dry soil samples.

No complete drying out of bacteria in dry soils was observed. Even in the cultures most sensitive to drying, there are single cells which are stable and survive for long periods in the dry state. Thanks to such cells the species does not die out under conditions of prolonged drought.

Great variations in the composition of the microflora also take place when the soil samples are kept moist. It is clear that the soil as a whole, and the separate soil aggregates possess different physicochemical conditions for the life of microbes than those present in isolated soil samples. Some bacterial species grow quicker in natural surroundings, others, slower.

Table: *Variations in the microflora in various samples of soil during moist storage (in thousands/g; counted on meat-peptone agar)*

Soils	*Bacteria, fresh*	*Bacteria, after 10 days*	*Myco-bacteria fresh*	*Mycob. after 10 days*	*Actino-mycetes fresh*	*Actino-mycetes after 10 days*
Sector 1, loam	23	40	0.5	8	0	0.5
Sector II, loam	40	56	0.8	22	0	0
Sector III, loose calcareous soil	91	150	1.5	65	0	1.0
Sector IV, loam	13	62	1.0	37	0	0.8
Sector V, loose calcareous soil	112	346	2.5	120	0	1.5

The total number of bacteria In the samples after 10 days increased 2 to 4 times, and the number of mycobacteria 16 to 50 times. Actinomycetes in fresh samples, were almost nonexistent, and after 10 days their numbers reached 500-1,500 per 1 gram of soil.

Similar changes in the composition of the microflora is noted in other soils during their storage in the moist state. In samples of podsol soil no *Azotobacter* could be detected by us after 2-3 days. The reason for this is the abundant growth of its antagonists—*Bac. subtilis* and *Bac. mesentericum.* In samples of chestnut soil mucolytic bacteria grew abundantly, and fungi of the genus *Fusarium* disappeared almost entirely.

For the determination of the microflora of the soil one has to take into account the composition of the medium into which the microorganisms are being inoculated. Experiments show that organisms from many soils grow better on synthetic media of Chapek, CPI, etc, than on media containing protein. On starvation media (water agar) and semistarvation media (Ashby agar) the number of bacteria is often 2-5 times greater than on rich nutrient media.

Table: *The number of soil bacteria after inoculation of various media (in thousands/g)*

Medium	***Garden soil***	***Primitive soil, mountainous, 3,800 m***	***Primitive soil, Severnaya Zemlya***
Meat-peptone agar (MPA)	3,500	54	23
Synthetic medium of Chapek	3,800	270	154
Synthetic medium CPI	4,400	850	—
Ashby medium	4,200	680	187
Water agar	3,800	800	—

It should be pointed out that it is easier to isolate bacteria from primitive soils,or mountainous soils (mountain summits, islands of the Arctic Ocean, etc) on starvation, semistarvation or synthetic media which do not contain protein. The bacterial colonies on such media are very small and can often be seen only with the aid of a a magnifying glass, or even only under a microscope. Such microcolonies usually consist of a few cells only.

The microflora detected after inoculation on starvation and semistarvation media differs, from that found in peptone media. The predominant organisms capable of growing on the synthetic medium of Chapek, CPI, etc are the auxotrophs (prototrophs), which do not require any growth factors or organic nitrogen. They can synthesize all the necessary biotic substances such as vitamins, auxins, etc.

On water agar and on the Ashby medium microorganisms usually grow at the expense of their food reserves. Among these organisms auxoautotrophs and auxoheterotrophe can be detected likewise. Often the so-called oligonitrophils grow on the nitrogen-less medium of Ashby. These are unique forms of bacteria and mycobacteria which are capable of nitrogen fixation in small quantities, satisfying their growth requirements (Mishustina, 1953).

On peptone media, and generally on media rich in organic substances, predominantly auxoheterotrophs (metatrophs) grow.

Auxoautotrophs also grow on these media. The quantitative ratio of prototrophs to metatrophs varies from soil to soil. In soils rich in humus, and well-fertilized with organic fertilizers, the amount of the former and the latter is approximately the same.

Organic, protein-containing media, are toxic for many soil microorganisms.

Apart from the afore-mentioned microorganisms, there are great numbers of organisms in the soil possessing specific functions. Such organisms can be detected on special, so-called selective media. To such bacteria belong the nitrifiers, sulfur bacteria, iron bacteria, *Azotobacter,* cellulose-decomposing bacteria and others. Special media are required for the cultivation of such bacteria.

The principle underlying the use of selective media (Vinogradskii, 1952) is as follows: in the selective medium, favorable conditions exist for the detection of a given function. It should be noticed, that the selective media are of relative importance. Investigations have shown, that many if not all prototroph bacteria have the capacity of growing on complex nonselective media. For example, *Azotobacter* can grow on nitrogen-less media due to the capacity of nitrogen fixation, but it can also grow on media containing inorganic and organic nitrogen.

Experiments have shown, that selective media are not strictly selective. No matter what the composition of the selective medium and how carefully it is prepared, other bacterial forms grow on it in addition to the desired bacteria.

On the nitrogen-less medium of Ashby or Beijerinck, apart from *Azotobacter,* many oligonitrophils and metatrophs grow well. On the medium of Vinogradskii, used for the nitrifiers, bacterial satellites also grow well. On media containing cellulose, not only bacteria capable of decomposing cellulose grow, but also other forms of bacteria.

Selective media are not optimal for the growth of bacteria. In. many cases, bacteria grow better upon addition to these media of ready sources of nutrition. Nitrogen compounds may be added for the growth of *Azotobacter,* sugar and other organic compounds for the growth of the cellulose-decomposing organisms, protein and nonprotein substances for others, etc (Rotmistrov, 1950).

According to Kalinenko (1953a, b) iron bacteria and nitrifiers grow well on ordinary organic and even protein-containing media.

It is evident that there are no strictly selective media. Universal media do not exist either.

The data presented in this chapter provide the basis for the assumption that the data on bacterial numbers in the soil are rather lower than in reality. Knowing their numbers, their total mass can be determined, or in other words, the soil productivity.

Cocci are 0.7 μ in diameter, their volume is 0.18 μ^3, and their weight 7 x 10^{-10} mg. About 5 x 10 ^{9} cells are present in 1 ml. The size of nonsporiferous bacteria is on the average 3 μ x O.7 μ, the cell volume 1.15 μ^3, and the weight 10 $^{-9}$ mg. About 900-1,000 million cells may be present in 1 ml. Cells of larger size (5 μ x 1 μ) have a volume of 3. 9 μ^3, their weight is 10-8 mg. In 1 ml there are about 350 million cells.

According to the data of Tanson (1948), in 1 ml there are 1,000 cocci of 1 μ in diameter; 330 million sporiferous bacteria of the size of 3 μ x 1 μ, and 1 million spores of fungi, 10 μ in diameter. According to Van Niel (1936),there can be 1,400 million cells of *Bact. coli* per 1 ml; according to Butkevich (1938), 10 ^{9} cells weigh 0.5 mg; Jensen (1940) found that 10 ^{9} cells of *Azotobacter* weigh about 5 mg. Similar data are given by Kendall (1928), Strugger (1948) and some other authors.

We have obtained the following data on the total microflora of the rhizosphere of vegetative plants. There are 2-2.5 kg of cells in a soil under lucerne in Central Asia, per 120 kg of soil; i.e., 6,000-7,000 kg of cells per hectare. Outside the root zone there are, according to our calculations, 1,500-2,000 kg bacterial cells per hectare of the upper (plow) layer. Consequently, there are about 7-9 tons of bacterial mass per hectare (Krasil'nikov, 1944).

In soils of medium fertility the total mass is considerably smaller. For example, in podsol soils under two-year clover and frequently fertilized we have found 1,000-3,000 millions of organisms per gram of soil in the rhizosphere and in the zone outside the roots, 300-800 million organisms per gram of soil. The total bacterial mass in the root zone amounted to 1,200-3,000 kg and outside the root zone about 350-1,000 kg. The total bacterial mass per hectare was 1,500-4,000 kg.

In the same soil under wheat, there were 800-1,200 million organisms per kg in the rhizosphere, and 100-200 million outside the roots. The total mass of bacteria was 1,100 kg per hectare.

In a poor. lightly cultivated soil (podsol) we have found under wheat, only 100-150 kg of bacterial mass per hectare in the upper (plow) layer. Eighty per cent of this mass was found in the rhizosphere.

Strugger (1948). on the basis of his investigations and those of Kendall, calculated that the total bacterial mass comprises 0.03-0.28% of the weight of the soil. Clark (1949) has shown that the bacteria constitutes 300-3,000 parts per million by weight of the soil. These data agree with our own.

Similar numbers are given by Khudyakov (1953c), Mishustin (1954), Berezova (1953) and others.

It should be recalled that our calculation takes into account only bacteria, whereas other organisms living in the soil, such an actinomycetes, fungi, algae, and protozoa are unaccounted for. They comprise a considerable mass of living substance.

The total number of fungi and actinomycetes per gram of soil runs to tens and hundred thousands, and not infrequently millions of organisms per gram of soil. The number of algae reaches thousands and hundred thousands and the diatomaceous algae 100 million per gram of soil (Brendemuhl 1949). The total mass of these organisms cannot be calculated owing to the peculiarities of their structure and growth. Nevertheless. according to the investigators, it is only slightly less than the total bacterial mass.

The total mass of protozoa and insects per hectare is 2-3 tons (Gilyarov, 1949, 1953).

The total mass of the living organisms does not represent merely a static reserve of organic substances, but a living active mass with a large potential, This mass is in constant growth. The individual cells of this mass grow, reproduce, grow old, and die. A constant change and regeneration of the whole living mass takes place.

Under natural soil conditions bacteria give on the average no less than two generations per month during the whole vegetation period, which lasts 7-9 months in the south and 3-5 months in the moderate belt. Consequently, the entire bacterial mass undergoes regeneration 14-18 times during the summer (in the southern belt), and 6-10 times in the moderate belt. The total bacterial production in the upper (plow) layer reaches tens of tons of living mass for one vegetative period.

The intensity of bacterial growth in the soil was determined by the time required for the doubling of their numbers. Three organisms were used in the experiment. *Az. chroococcum, Ps. aurantiaca* and *Bact. prodigiosum*. A sample of a garden soil was placed in an asbestos bag inoculated with the above-listed organisms, carefully mixed and immediately subjected to a microbiological analysis. The soil in the bag was washed with water, until all the desorbed bacteria were removed, and then the bag was buried in the same garden soil from which the samples were taken. After 1-2 days the bags were taken out and the soil was subjected to the same procedure as before. This was repeated for a month. The experiments were carried out in May, July-August and September-October, three series in all. In each series 100 million organiams,were introduced into the soil. In the first analysis (immediately after mixing) 26 million *Azotobacter* cells were washed out. 18 million cells of *Ps. aurantiac*a and 34 million cells of *Bact. prodigiosum* (all these numbers are per gram of soil). The rest of the cells were in the adsorbed state, but they did not lose their capacity to reproduce.

Upon subsequent analyses the following amount of cells was washed out (the May experiment), in millions/ g:

Analysis	***Az. chroococcum***	***Ps. aurantiaca***	***Bact. prodigiosum***
Second	16	22	14
Third	23	26	10
Fourth	20	28	8
Fifth	24	34	10

As can be seen from the above data, the doubling of *Azotobacter* cells took place every 5 days, *Ps. aurantiaca* every 4 days. and *Bact. prodigiosum* every 10 days. In other words the number of generations of the first was 6, of the second, 7 and the third 3. In July-August the number of generations was 4, 4 and 2 respectively, and in September-October, 4, 3, 1 respectively.

The results of these experiments served as a basis for the calculation of the speed of growth of the bacterial maps in the soil.

Distribution of Microorganisms in Soil

The problem of microbial distribution in the soil is scarcely dealt with in the literature. We know almost nothing about the localization of microorganism in the soil. It is generally assumed that microbial cells are uniformly distributed in the soil penetrating all pores by diffusion. Therefore, upon quantitative calculation of seperate groups and bacterial species one usually limits himself to one or two soil samples.

Such an assumption does not conform to reality, and the data so obtained lead to erroneous conclusions as to the distribution of the individual microbial species in the soil,

Our studies show that the distribution of microbial cells in soil is not diffuse but focal. In each focus large or small cells of one species or several nonantagonistic species, grow and concentrate. Microbes, especially bacteria and mycobacteria inhabit soils in colonies (Krasil'nikov, 1936).

The focal character of microbial distribution in the soil is being confirmed by the daily practice of microbiological soil studies. It to known that in one and the same field, *Azotobacter,* for example, may be found in one sample and not in another. If one takes 100-200 samples from 1 hectare of a given soil, cells of *Azotobacter* will not be found in all samples taken, depending upon the numbers of *Azotobacter* in this soil. The latter is determined by the state of the soil. In fertile, well-cultivated soils, rich in humus, *Azotobacter* will be found in every sample. In poor, nonfertile or virgin soils *Azotobacter* is rarely encountered in all the samples.

Azotobacter was detected by us In all 200 samples taken from 1 hectare of a cultivated serozem soil under 3-year lucerne in Central Asia. In virgin soils we have found this microbe only in 3 samples (out of 2 00) and in poorly-cultivated soils in 45-85 samples. Similar results were obtained when studying the soils of the Volga area (chestnut soil and others) and the podsol soils.

The soils of the Moscow Oblast' which were under forest until 5-10 years previously and now under various plants did not contain *Azotobacter*. We did not detect *Azotobacter* cells in any of the 1,250 samples studied. In garden soil this microbe was found in almost all samples (400 samples studied). In field soils (well -cultivated) *Azotobacter* was found in 60% of samples (860 samples were studied).

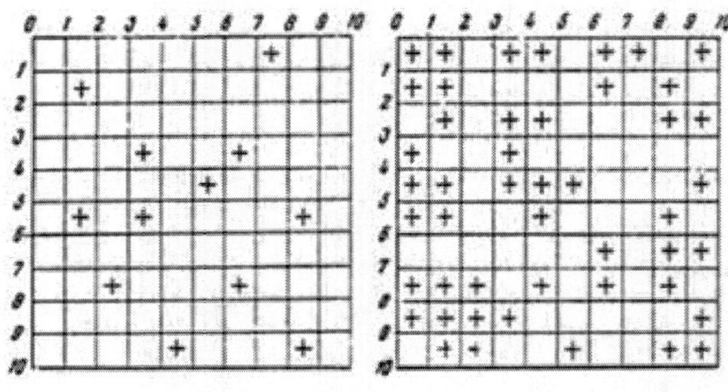

Figure: *Schematic representation of the character of distribution of* Azotobacter *in soil. The sign "+" shows the presence of* Azotobacter *in 1 square meter soil sectors a) poorly cultivated soil; b) well-cultivated soil (podsol, Moscow Oblast'). 1-10-numeration of squares.*

For a more accurate determination of the focal distribution of *Azotobacter* in the soil we have carried out the following experiment (in the experimental stations near Moscow). One-hectare plots in three fields containing different numbers of *Azotobacter* were studied.

On these hectare plots, 3 one-meter sectors, along the diagonal, were thoroughly analyzed for the presence of *Azotobacter*. To this end these sectors were divided into 100 small squares and 15-20 grams of soil were taken from each square. A total of 300 samples were analyzed from each hectare plot. The results of these analyses are given, and the plan of *Azotobacter* distribution in the one-square-meter sectors.

Table: *Distribution of* Azotobacter *in the soil*
(number of samples containing Azotobacter *in %)*

Sectors in fields	***Sector I***	***Sector II***	***Sector III***	***Average***
Garden soil	92	90	96	92.7
Field, well-cultivated, under 3-year clover	53	60	34	49
Field, poorly cultivated, under 3-year clover	16	8	12	12

These data show that even in garden soil well-cultivated and systematically fertilized with mineral and organic fertilizers, *Azotobacter* cannot be detected in every sample tested. Of 100 samples taken from the 1-square-meter sectors, it was found in 90-96 samples. In field soils, well-cultivated and fertilized the number of samples containing *Azotobacter* was 34-60 and in poorly-cultivated soils, as well as in soils only recently under cultivation (5-10 years of cultivation, previously under forest) it was found in 8-16 samples out of 100.

More detailed examination of the soil sample reveals small foci in which the *Azotobacter* cells are located. It is known from laboratory practice that when the soil sample is placed in small lumps on Ashby agar or gel plates, not each lump gives colonies of *Azotobacter*.

The percentage of growth from each lump varies from 0%-100, depending on the soil. The method of placing soil lumps for the detection of *Azotobacter* and other species of bacteria (nitrifiers, cellulose decomposers, etc) is extensively used in microbiological practice.

We have analyzed samples of the well-cultivated soil of the same field as was the object of our previous experiments. Samples were taken from square-meter sectors, in the form of monoliths.

Each sample after thorough mixing was divided into 5 samples of 0.1 g weight. Each such sample was divided into 100 small lumps, which were placed on the surface of Ashby agar (in Petri dishes). Five samples were taken from each of the sectors studied. The results are given.

Table: Azotobacter *distribution in soil samples (growth from lumps, %)*

No .of sample	*Sector I*	*Sector II*	*Sector III*
1	23	70	28
2	99	39	15
3	97	55	31
4	88	86	8
5	56	42	21
Average	72.6	58.4	20.6

The data presented in Table 27, show that the distribution of *Azotobacter* even in small lumps of soil is focal. The number of microfoci containing *Azotobacter* in such a lump is determined by the total number of *Azotobacter* in such soil and by other factors. In the samples studied by us, the lumps of 0.1 g contained in some cases from 21-99 and in others from 0 to 3-5 microfoci where *Azotobacter* could be detected.

It should be noted that these foci are so small that they are not destroyed during ordinary mechanical crushing of the soil samples. In our experiments, we have carefully mlxed the soil samples and in spite of this, uniform distribution of *Azotobacter* was not achieved. The microfoci were thereby not destroyed or only a small fraction of them was destroyed. The percentage of lumps containing *Azotobacter* was almost identical to that of samples not subjected to mechanical crushing. In crushed samples 50-60 microfoci were found and in the intact noncrushed 30-50. Only when the lumps were ground into dust were the foci destroyed.

The focal distribution and *Azotobacter* cell concentration described is also characteristic of other bacteria. It is well-defined in nitrifiers, cellulose decomposers, root nodule-bacteria, mycobacteria and others. It is less well-defined in fungi, and actinomycetes as a resuit of their biological pecularities. Rippel-Baldes (1952) noted focal development of *Aspergillus niger* in the soil. In square-meter sectors he found this fungus only in 14 squares out of a total of 100.

The focal concentration of microbial cells in soils can directly be seen under the microscope, employing the method of Rossi-Cholodny, To this end cover glasses are immersed in the soil and after some period their surface is overgrown with microbes, bacteria, actinomycetes, fungi, yeasts and others.

We have found colonies consisting of several cells of a size not greater than 10 μ, Clusters of microorganisms can be encountered which occupy an area of a 100 μ cross section. More frequently, colonies of moderate size (20-70 μ in diameter) and consisting of several tens of cells are encountered.

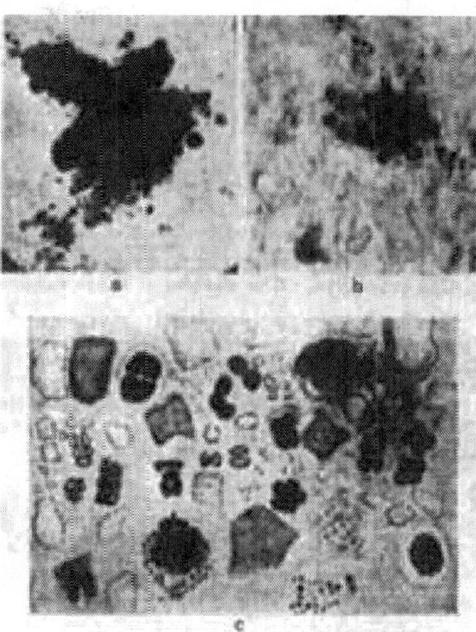

Figure: *Distribution of microorganisms in soil a) large colonies of* Azotobacter; *b) small colonies; c) general view on the distribution of microbes in the preparation (according to Vinogradskii, 1952).*

In our experiments we have noted the formation of colonies of *Azotobbacter,* sporiferous and nonsporiferous bacteria. Frequently, formation of colonies of mycobacteria, proactinomycetes and actinomycetes could be observed. Actinomycetes thereby form conidia with well-developed spores. Not infrequently actinomycetal hyphae develop into rodlike and coccoid cells, in the same way as can be observed on artificial nutrient media.

Branches with sporeforming cells are seen (spiral and straight). Many threads disintegrate into rods and cocci as in proactinomycetes (microphotos from glass imprints according to the method of Cholodny).

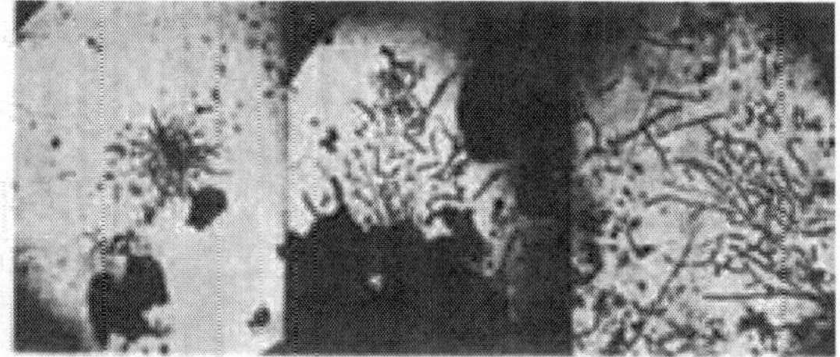

Figure: *Colonies of actinomycetes in the soil*

The frequency of colony formation on cover glasses depends upon the soil properties. Especially great numbers of colonies are formed in the rhizosphere of plants. Bacteria and mycobacteria grow around thin root tips and around root hairs either in confluent layers as was noted earlier, or in separate foci, in formless clusters or in colonies.

The formation of colonies in the soil has also been described by Cholodny (1934), Kubiena (1932), Rossi (1936), Vinogradskii (1952) and others. The microphotos given by us show clearly enough the concentrations of bacteria, fungi and actinomycetes.

It is self-evident that together with the colonies on the cover glass, individual cells can be seen. They result from the destruction of the integrity of the colonies. The picture of microbial distribution in the soil is clearly seen under the microscope in ultraviolet light, especially after staining with acridine- orange. The individual cells and colonies of bacteria, actinomycetes and fungi standout sharply, glowing with a green color. Only a few cells glow with a red color. We have studied different soils according to this method, and always obtained positive results.

In those cases when pores contain air, colonies of fungi and actinomycetes form fruit-bearing hyphae. In large pores, under favorable conditions, microbes proliferate and are concentrated in larger numbers than in the small pores. Bacteria, fungi and actinomycetes can penetrate adjacent pores through capillaries.

Some investigators assume that microbes do not penetrate the very small pores of the soil. It was shown above that there exist organisms of ultramicroscopic dimensions lying on the border or beyond the border of visibility in optical microscopes (0.05-0.1 μ). To these belong phages (bacteriophages, actinophages), filterable bacterial forms, then certain cellular elements the so called L-forms, special regenerative bodies, etc of the ordinary species of bacteria and actinomycetes, and finally, small cells of individual organisms.

Investigations showed that not only ultramicroscopic organisms can penetrate through the pores and small capillaries but also many bacteria and actinomycetes of normal size, of a diameter of 0.3-0.7 μ and more.

It is known that bacteria and actinomycetes do not pass through filter candles upon ordinary filtration under pressure. Sterilization by filters is based on this observation., In laboratory practice the most widely used filters are the Berkefield filter (L_3 L_5), Chamberlain filter (N); Seitz filters or membrane filters with pores of a definite size.

If such filters are filled with liquid bacterial suspension and immersed in a nutrient solution (meat-peptone broth), then after an incubation period at 25-37° C, cells will pass through their walls and start growing outside them. In our experiments the following microorganisms passed through such filters: *Bact. coli, Bact. prodigiosum,.Ps. aurantiaca,* larger sporiferous species, *Bac. mycoides, Bac. mesentericus, Bac. megatherium,* actinomycetes. *A.violaceus, A. coelicolor,* and *A. globisporus.*

Besides bacteria, actinomycetes and fungi in the course of growth pass comparatively easily through small-pore clays. We studied the clays of natural deposits underlying the soils of the Moscow district fields. Clay samples were placed in Koch dishes, wetted with water uniformly mixed and distributed in a layer of 1.5-2.0 cm. Bacteria were introduced into the center well. After incubation at 25° or 37° C samples, from different distances from the center well, were taken at various time intervals and analyzed. The experiments were carried out in sterile and nonsterile conditions. In nonsterile conditions easily detectable microbial species were employed: *Bact. prodigiosum, Ps. aurantiaca, Bact. coli, Az. chroococcum, Bac. Mycoides, A. violaceus.* In sterile experiments, besides the afore-mentioned *Bac. mesentericus, Ps. fluorescens* was also employed.

As can be seen from the data, the speed of growth and bacterial mobility is not the same in the different bacteria and actinomycetes and varies in relation to the properties of the clay. The threads of actinomycetal mycelium move with the greatest velocity. Some nonsporiferous bacteria grow rapidly. Cells of *Bact. coli* grow slowly and *Bac. mycoides* is the slowest of them all.

The microbial cells do not move through small pores of the natural and artificial substrates mechanically, or under the action of external pressure, as they do under filtration, but they move as a result of overgrowth. Dead cells do not pass through filters. There is no movement of cells, or only very weakly, when filters with living bacteria are immersed in pure water.

The more intense the growth of microbes, the more rapidly they pass through small pores and capillaries. The optimal temperature for the growth and multiplication of bacteria Is, at the same time, the most favorable for their passage through small pores. At a temperature of 5-7° C *Bact. coli* and *Bact. prodigiosum* multiply slowly and pass through a Chamberlain candle in 60-80 hours. At a temperature of 25° C this period is shortened to 20-30 hours. The motility of microbes in clay is increased upon introduction of organic substances into the clay as meat- peptone broth, or saccharose. To do this, a well is made in the clay which is prepared in the same way as in the preceding experiment, and, at some distance from it, an elongated groove is dug. Bacteria are placed in the central well (aqueous suspension), and nutrient substances (those listed above) in the groove.

It was noted that microbial cultures of *Bac. mycoides, Bact. prodigiosum, Ps. aurantiaca,* and *A. coelicolor* moved in the direction of the nutrient broth quicker than in the control experiments. In one day *Bact. prodigiosum* moved, in the control experiment, 1.5-2.0 cm, in the presence of the meat-peptone broth 3.0-3.5 cm, in the presence of saccharose 2.5-3.0 cm. *Bac. mycoides* in the control moved 0.7 cm and in MPB 1.5 cm; *Ps. aurantiaca* moved 2.0 cm in the control as compared with 3.0-4.0 cm in the experiment. The movement of actinomycetes in the presence of organic substances was by 1.0-1.5 cm greater than in the control. It should also be pointed out that the degree of permeability of small porous substrates also depends on a great number of external factors, such as the pH of the medium, aeration, and the composition of the soil solution. Anything that favors the growth and proliferation of organisms also enhances the overgrowth of the substrate.

Thus, the data obtained established the possibility of penetration of microbial cells through the smallest soil pores, under natural conditions.

Apparently no soil pores exist which cannot be penetrated by microbes.

Chapter 5

Biotic Substances of the Soil

It can be seen from the above data that small doses of organic substances, present in processed compost or manure, are needed for the improvement of the growth of plants. The question arises which of these organic substances are the activators of plant growth. It was said above that the organic part of the soil consists of a multitude of complex compounds with specific and nonspecific proportion. The majority of these compounds are related to humic acids. Beside the humic acids many other substances were found in the organic part of the soil. They include substances with biocatalytic properties, Among these are enzymes, vitamins, auxins, certain amino acids and other biotic substances. Their action is similar to that of the yeast bios. The analysis of peat infected with bacteria showed the presence of purine bases. The latter were also found in the cells of *Azotobacter*. This was the reason for Mockeridge's conclusion that the activating compound of composts and *Azotobacter* to one and the same substance.

It wax shown in later studies that the action of humus is caused by special substances produced by microbes (Nath, 1932, McCarrison, 1924, And others). These substances be called phytamins. According to him the phytamins are formed by microorganisms only. Upon entering the plants they are transformed into vitamins. The latter entering with food into the body of animals and human beings are subjected to certain transformations and are converted into hormones. Hormones and vitamins are excreted from the animal body with urine or excrements, find their way into the soil or manure, And are there transformed by microorganisms into phytamins.

In such a way, according to Anstead, phytamins, vitamins and hormones are products of one and the same substance. Analogous views were hold by Daineko (1939). He assumed that there is no difference between the animal hormones and vitamins, and that they are interlinked. The link between plant and animal biocatalysts is stressed by many inventigators. However, there are no grounds for speaking of any cycle of these substances in nature.

Lochhead and Chase (1943) attempted to elucidate the nature of the activating substance of humus. They prepared extracts from soil humus and subjected it to various procedures, such as extraction with organic solvents, absorption with carbon, and other absorbents followed by elution. The experiments showed that the ash of humus and composts had no effect whatsoever on the growth of plants and microorganism, Of 63 microbial cultures, only one grew in the presence of ash obtained from composts. The extracts obtained from composts and humus of the soil, however, had a positive effect on their growth.

The acetone extract was the one most effective, In its presence 26 cultures grew well, 26 cultures adequately, and only 11 cultures did not grow. Alcohol and ether extracts also gave good results, Filtrates and eluates from charcoal, each gave a small effect, their effect was larger if they were combined. Allison and Hoover (1936) noted a positive effect of humic acids on the growth and activity of root-nodule bacteria. They have discovered the presence in humin of a special substance which they called "Factor R" or "Coenzyme." Robinson and Endington (1946) found a biotic substance in soils "fluorin." This substance, according to them, is absorbed in considerable amounts by plants.

The quantity of auxins varied according to the soil properties. In manure-fertilized soils the amount of auxins was 0. 200 μ g/ kg and in poor, nonfertilized soils only 0.06 g/ kg of soil. These substances are found in lesser amounts (0.09-0.106 μ g/ kg) in soils under mineral fertilizers. The soil is enriched in auxins in the course of composting. Sterilized soil contained 0.042 μ g auxins per kg of soil and after 6 days incubation in a glass house, under favorable temperature and humidity conditions, they increased to 0.146 μ g/ kg of soil. Williams, Stewart, Kejes and Anderson (1942) found larger quantities of auxins in soils than did Parker-Rodes. According to their data, the A-horizon of fertile soils contains up to 175 μ g auxins per kg of soil, Nonfertile soils contain 40-60 μ g auxins per kg of soil. In horizon B auxins were not detected.

Hamense (1946) by the use of improved methods of analysis found 100 times more auxins in soils than did previous workers. According to his data the auxin content was 160-450 μ g per kg of soil (Schmidt, 1951, and Hamense, 1946). Stewart and Anderson (1942) had shown that the amount of auxins in fertile soils is entirely sufficient for the stimulation of plant growth. Hamense (1946) found in different soils 0.16-0.045 μ g equivalents of ß-indoleacetic acid in 1 g of dry soil. The auxin content increases after the application of organic fertilizers and then falls below the initial level and afterward rises again to the normal level, characteristic of the given soil. According to the author's observations, chemically pure preparations of auxins are not preserved in the soil for long periods.

According to Matskov (1954), chemical preparations of 2. 4-dichlorophenoxyacetic acid are preserved in the soil for the duration of a whole winter (5-6 months). Another growth stimulant—heteroauxin (ß-indoleacetic acid) is not preserved in the soil that long.

Auxins were also detected in the soil by Roberts I. and Roberts E. (1939) and some other investigators.

Beside auxins, soils also contain many other biotic substances such as vitamins, biotin, nicotinic acid, pantothenic acid, folic acid, amino acids, various growth factors R, Z, X, and others, These substances have been found in the following quantities:

Thiamine	0.29-1.93 μ g	Roulte and Schopfer, 1950; Schopfer, 1943
Riboflaven	9.0-980 μ g	Schmidt and Starkey, 1951; Carpenter, 1943
Biotin	23.0-62.0 μ g	Roulet and Schopfer, 1950
Vitamin B_6	amounts not indicated	
Vitamin B_{12}	0.2-1.5 μ g	Robbins, Hervey and Stebbing, 1951, 1952
Inositol	amounts not indicated	
Nicotinic acid	amounts not indicated	Roulet, 1948
Para-aminobenzoic acid	amounts not indicated	Roulet, 1948
Pantothenic acid	amounts not indicated	Roulet, 1948
Folic acid	amounts not indicated	Roulet, 1948
Factor X	amounts not indicated	Lochhead and Texton, 1950
Unknown factor	amounts not indicated	Lochhead and Texton, 1950

Apart from complete vitamin molecules, molecule fractions are encountered in nature which have the same effect on some organisms as the whole molecules, for example, the pyrimidine, thiazole and others (Lilly and Leonian, 1939).

We have found vitamins in different soils; as a rule, they were in larger quantities in places where the microbiological processes were more intense. Thus, they appear in chernozems in larger amounts than in podsol soils.

Table: *Vitamin and bacterial content of various soils*

Soils	***Riboflavin μ g/100 gm***	***Thiamine μ g/100 gm***	***Biotin μ g/100 gm***	***Bacteria millions/ gram***
Chernozem (Moldavian SSR)	98.0	4.5	45.0	1,500
Podsol(Moscow Oblast')	5.0	1.2	25.0	0.5

Cultivated soils contain more vitamins than virgin soils.

The vitamin content of the soil is qualified not only by the extent of cultivation but also by the nature of the vegetative cover. Plants which favor abundant growth of microorganisms, as a rule, assimilate more biotic substances in the soil. In the serozems of Central Asia the highest concentrations of biotic substances were found under 2-3-year-old lucerne, their concentration was lower under cotton and very little was found in virgin soils.

Table: *Vitamin and microbe content of the soils of Central Asia, under various plants*

Soil	***Thiamine μ g/100 gm***	***Biotin μ g/100 gm***	***Microorganisms millions/gram***
Virgin soil, the valley of the river Vakhsh	1.5	10.0	0.5
Virgin soil	0	+	0.1
Cultivated soil (2 year-old lucerne)	6.5	38.0	4,500
Cultivated soil (cotton long under cultivation)	3.0	18.0	1,500

According to Shavlovskii (1954, 1955), soil under potatoes contains 0.5 μ g/kg biotin, and soil under clover—l.3 μ g/kg. In the Lvov Oblast' in serozem forest soils and podsol chernozem the author obtained the data shown.

Table: *Vitamin content of soils under various plants, in μ g/kg*

Soils	***Biotin***	***Riboflavin***	***Nicotinic acid***
Serozem forest, under wheat	0.3	4.0	100.0
The same, under grasses	0.7	7.0	230.0
Podsol chernozem, under wheat	0.8	10.0	280.0
The same, under grasses	1.5	14.0	350.6

In soil adjoining the root system there are more biotic substances than outside the rhizosphere.

In one kg of soil from the rhizosphere of wheat grown in fertilized fields of the Experimental Station of Dolgoprudnoe (Moscow Oblast') we found per 100 g soil 10 μ g of thiamine, 150 μ g riboflavin, 35 μ g biotin; outside the rhizosphere, 1.2 μ g thiamine, 25 μ g riboflavin, 3 μ g biotin; in the rhizosphere of tobacco, 10-15 μ g thiamine, and outside the root zone 1.5-4.0 μ g per 100 g of soil.

According to Shavlovskii, the amount of vitamins in the rhizosphere of buckwheat is twice as high as that outside the root zone. On the 20th day of growth he found:

	in the rhizosphere, ***μ g / kg***	***outside the rootzone*** ***μ g / kg***
Nicotinic acid	600	260
Biotin	2	0.5
Vitamin B_6	8	—

Higher concentrations of vitamins in the rhizosphere were noted by Roulet (1954). The amount of biotic substances in the upper layer of the soil is higher than in the lower layers. The highest concentrations are found in the upper layer (0-20 cm).

Table: *The distribution of vitamins in soil layers (μ g/kg)*

Soil	***Layer, cm***	***Thiamine***	***Biotin***
Podsol of Moscow Oblast			
under 2 year clover	0-20	2.3	14
	20-30	0	3.0
	50-70	1.9	8.0
	80-100	0	0
Serozem of Vakhsh Valley			
under 3 year lucerne	0-20	5.6	42
.	30-40	1.8	14
	45-60	2.4	21
	70-90	1.1	4.2

In some soils a small increase in the concentration of biotic substances is noted at a depth of 60-70 cm.

Some other investigators have noted the decrease in concentration of vitamins in deep layers of the soil. Roulet and Schopfer (1950) give the following data on the distribution of vitamins according to, layers (per 100 g of soil):

layer, cm	*thiamine, μ g*	*biotin, μ g*
10	1.93	62
20	0.86	39
30	0.62	27
50	0.29	23

These authors found thiamine and biotin even at a depth of 2.0 to 8.5 m.

Lilly and Leonian (1939) found considerable quantities of thiamine and biotin, as well as vitamin B_1 and its components thiazole and pyrimidine, in the soil. They were concentrated in the upper layer of the soil. At a depth of 60 cm they were not detected. West and Wilson (1938, 1939) found thiamine and biotin in the root zone of tobacco and flax. According to Schmidt and Starkey (1951), riboflavin can be found in the soil in varying amounts, according to soil fertility, vegetative cover, etc. In soils under forests the authors found 500 μ g of riboflavin and in soils under plow and in fertile soils about 10 μ g per 100 g of soil. After the application to the soil of organic fertilizers such as straw, grass or sugar, the amount of riboflavin increased. The more organic substances applied, the higher was the concentration of riboflavin in the soil. For example, after application of grass 15%*, about200 μ g riboflavin were found; if the amount of grass added was 10% 120 μ g riboflavin were found, in the presence of grass of 5% only 60 μ g riboflavin was found. The amounts of riboflavin are given per 100 g of soil.

By the end of the growth period, in August-October, more vitamins are found in the soil than in the spring. The amount of riboflavin found in spring was 80-300 μ g, in autumn it reached 600-980 μ g per 100 g of soil.

The amount of biotic substances in the soil constantly changes, according to external conditions such as temperature, humidity, season, etc. The vitamins are preserved in the soil for various periods of time, the length of which also depends upon soil and climatic conditions. According to Stewart and Anderson (1942), growth-stimulating substances can persist in dry soil for 3-4 years, According to Schmidt and Starkey, 50% riboflavin can be detected in fresh soil after 3 days, pantothenic acid is completely decomposed in one day.

Some amino acids can be listed among the growth factors of lower and higher plants. Plants synthesize these amino acids in the same way they synthesize vitamins, but, nevertheless, the addition of small doses of amino acids has a positive effect on the growth of plants. Nielsen and Hartelius (1938) tested amino compounds, Six of them—ß-alanine, asparagine, aspartic acid, glutamic acid, lysine and arginine markedly enhanced the growth of lower organisms; ß—alanine had the greatest effect. Five μ g of ß-alanine per 50 ml of medium was sufficient to enhance the growth and increase the yield by 66% dry weight. The maximal stimulation was found at the concentration of 1:100,000. Arginine acts at a dilution of 1:20,000 and lysine at a dilution of 1:4,000. Glutamine exerts a positive effect on the growth of organisms in a dose of 1:1,000. Substances in such doses can be considered as sources of nutrition. Other amino acids such as asparagine and others act at even higher concentrations and cannot be regarded as growth factors but as nutrients.

The amount of amino acids in soils varies; it depends upon the properties of the latter and upon climatic conditions. The more fertile the soil, the more amino acids it contains. The concentration of amino acids is determined by the rate of their influx into the soil and by the length of time of their preservation. This list does not include all the biotic substances of the soil. It should be assumed that the soil, the same as other natural substrates, contains many other substances unknown to us, which act as biocatalysts, enhancing the metabolism of organisms. Lochhead and Texton (1940, 1950, 1952) detected activating substances in the soil which were of an unknown nature and could not be replaced by any of the known growth factors.

The Origin of the Biotic Substances of Soil

All the diverse biotic substances have their origin in the metabolic activity of plants and microbes.

Formation of biotic substances by the plants. It was mentioned above that, during their lifetime, the roots of many plants excrete substances which stimulate the growth of organisms, For example, the seeds of broomrape germinate, only in the vicinity of roots of sunflower, flax, corn, soy and some other plants, The root excrements of these plants activate the growth of broomrape sprouts. The activating factor is thermostable and is, not decomposed on boiling and prolonged drying (Bartsinakii, 1935; Beilin, 1941).

Morning glory stimulates the germination of melon seeds. Pollen grains of angiosperms mutually stimulate each other on germination by excreting activating substances (Golubinskii, 1946). Found biotin and thiamine in root secretions of corn and peas. According to him, the more of these substances are secreted, the more intense is the growth of the plants. Corn secretes more biotin, and peas more thiamine. The root excretions contain vitamins in the following amounts per g dry weight of the plants: corn—0.5402 μ g of thiamine and 0.2308 μ g of biotin; peas—0.6634 μ g of thiamine and 0.2658 μ g biotin.

According to our observations, peas, wheat, and corn secrete more biotic substances in their early growth period than in the period of fruiting.

Table: *The presence of vitamins in root secretions of various plants (μ g per ml of nutrient solution)*

Plants	*After 10 days growth: Thiamine*	*After 10 days growth: Biotin*	*After 45 days growth: Thiamine*	*After 45 days growth: Biotin*
Wheat	0.1	0.6	0	0.1
Corn	0.2	0.5	0	0
Peas	0.5	1.5	0.1	0.7

Plants grown in aerated solutions secrete more biotic substances than those grown in poorly aerated solutions. For example, wheat grown with good aeration secreted 0.21 μ g thiamine, and wheat grown in the presence of small amounts of oxygen secreted 0.11 μ g thiamine. The respective amounts of biotin secreted were 0.8 and 0.6 μ g (per cm^3 of medium).

There are indications in the literature that germinating seeds of various plants secrete small amounts of vitamins into the medium (Meisel, 1950; Schopfer, 1943).

Biotic compounds can find their way into the soil together with decomposing residues. It is known that plants contain considerable amounts of various compounds which can stimulate the growth and development of organisms. For example, according to Burcholder and others (1944), the following amounts of thiamine are present in the tissues of: soya 47-61 x 10^{-7}; barley 28-51.8 x 10^{-7}; corn 17.6-31.2 x 10^{-7} (in mols per kg dry weight). The aerial parts contain more thiamine than the roots: in the leaves of corn were observed 17.6-30. 2 x 10^{-7}; and in the roots—4.06-9.78 x 10^{-7} mols per 1 kg of dry mass.

Auxins, vitamins of the B-group, bios, vitamins, D, K, C, H and P, pantothenic acid, paraaminobenzoic acid, nicotinic acid, purine derivatives and various hormones, etc, have been found in plants (Zeding, 1955; Ovcharov, 1955, and others.

Many biotic substances are excreted in the feces of animals and human beings. For example, human beings excrete daily in urine 60 μ g of thiamine, 600 μ g riboflavin, 626 μ g inositol and large quantities of pantothenic acid, as well as other compounds (Meisel, 1950). Vitamins of the B-group, pantothenic acid, nicotinic acid, folic acid, and other substances are found in the faces of human beings and animals.

The investigations show that these substances are synthesized by the microflora of the intestines (Perets, Gryazno and Agibalova, 1948; Nepomnyashchaya, 1950; Najjar and others, 1943-1950; Perets, 1955), The vitamins and other substances enter the soil with manure. According to Bonner and others (1938), 1kg of manure contains 130 μ g thiamine. Riboflavin, pantothenic acid, nicotinic acid, and other activating substances are present in manure. Sauerland (1948) found considerable amounts of substances of the bios type in manure, feces and urine of cattle, The quantities of these substances fluctuate with the seasons. The largest amounts being found in summer, less in winter. The amount of growth factors in manure and animal feces also varies according to the quality of the fodder.

All these substances accidentally find their way into the soil, and on the whole they constitute only a small traction of the total amount of biotic substances in the soil. The microorganisms are the main factor in the enrichment of the soil with these substances.

Microorganisms Synthesizing Biotic Substances

The capacity of microorganisms to synthesize biotic substances has been known for a long time,

Wildiers (1901) showed the presence of activating substances in yeast cultures, These substances were called by the author bios substances, Fifteen to twenty years later the attention of various specialists was drawn to these substances, They were found in various organic substrates, in plant tissues and in cultures of many microbes.

Investigations showed that biotic substances are synthesized by various microorganisms—bacteria, fungi, yeasts, actinomycetes and others (Meisel, 1950, Ierusaimskii, 1940 Kudryashov, 1948; Bukin, 1940, Stephenson. 1951, Schopfer, 1943).

The organisms are divided according to their capacity to synthesize biotic substances into auxoautotrophs and auxoheterotrophs. The former synthesize all the compounds necessary for their growth and can therefore grow on synthetic, vitaminless media; the latter do not synthesize, or more precisely, they do not synthesize all the necessary biotic substances and therefore they cannot grow on vitaminless media,

The capacity to synthesize various growth factors is present in many if not in all species of soil bacteria.

Chemosynthetic bacteria are the most active in synthesizing biotic substances, They can grow in purely mineral, completely vitaminless media, being capable of synthesizing organic substances from atmospheric CO_2. For example, thiamine, riboflavin, pantothenic acid, nicotinic acid, vitamin Be and some other compounds were found in the culture of *Thiobact. thioxidans* (O'Kane, 1943). These bacteria and other similar chemosynthetic bacteria—the nitrifters and hydrogen- and methane-oxidizing bacteria, synthesize their full quota of biotic substances. Without this ability they could not have developed in mineral media, Bacteria incapable of CO_2 assimilation, but growing well in synthetic vitaminless media with organic carbon sources, are also able to synthesize biotic substances, To such bacteria belong the bulk of the soil microflora *Azotobacter*, root-nodule bacteria, representatives of the genus *Peoudomonas, Bacterium,* oligonitrophils, mycobacteria, and others.

Boysen-Jonsen (1931) found that hetroauxin is synthesized by 16 bacterial species among which were *Ps. radiobacter, Bact. denitrificans, Bac. mycoides, Bac. subtilis, Bacterium* sp., and others. Raznitsyna (1938) employing the coleoptile method has detected the formation of this compound by various representatives of bacteria and mycobacteria. She divided microorganisms into three groups, according to their ability to synthesize auxins: a) organisms which do not synthesize them (or synthesize them in small quantities) *Mycobacterium rubrum, Az. vinelandii, Bac. mycoides,* and others, b) bacteria of medium activity —*Az. agile, Az. chroococcum* strains 31, 35, *Bact. coli, Myob. luteum, Ps. flourescens,* strains F. 24 and others, c) bacteria synthesizing large amounts of auxins—*Bact. proteus, Ps. fluorescens* strain 21, *Az. chroococcum* strain 54, *Mycob, album,* and others.

Roberts I, and Roberts E. (1939) studied the capacity of bacteria, fungi and actinomycetes to synthesize heteroauxins. This compound was synthesized by 99 cultures out of a total 150 studied, According to the authors, the most active producers of heteroauxins were bacteria and actinomycetes. Heteroauxin was found in different species of nonsporeforming bacteria—*Az. chroococcum, Pseudomonas, Bacterium, Vibrio, Mycobacterium,* and many other bacteria,

Beside auxins, a number of other biotic compounds such as thiamine, riboflavin, para-aminobonzoic acid, nicotinic acid, pantothenic acid, folio acid, vitamin C, vitamin K, vitamin B_3, vitamin B_{12}, inositol, biotin, provitamin D_2, alpha- and ßcarotenes, factor R, factor Z, and others can be detected in bacterial cultures (Detinova, 1937, Leo and Burris, 1943, Jones and Grooves, 1943; Burton and Lochhead, 1951,

Lochhead, 1952; Lochhead and Burton, 1955), The ability of the root-nodule bacteria to synthesize thiamine, riboflavin, pantothenic acid, vitamin B_{12} and others was discovered by Burton and Lochhead (1951), Went and Wilson (1938, 1939). These and other vitamins were found in cultures of *Az. chroococcum,* in various species and strains of the genus *Pseudomonas* and in mycobacteria. In recent years vitamin B_{12} was found in many bacteria and especially in actinomycetes, Some of these organisms such as *Mycob. propionicum, A. rimosus, A. aureofaciens,* and others, are employed in industry for the production of this vitamin. According to our observations, 90-95% of the actinomycetes isolated from soil synthesiss vitamin B_{12} (Krasil'nikov, 1954 c), According to Darken, 64-66% of soil actinomycetes synthesize this vitamin, (Darken, 1953).

Yamagutschi and Usami (1930) found vitamin B_2 in 15 bacterial cultures *(Bac. subtilis, Bac. Mesentericus, Bac. mycoides, Bact. prodigiosum,* various micrococci, and others).

Landy, Larkum and Oswald (1943) found paraminobenzoic acid in cultures of 35 species of bacteria and actinomycetes. Among these were: *Bact. proteus, Ps. pyogenes, Bac. aerogenes, Bac. subtilis, Bac. megatherium, Mycob. diptheriae, Mycob. stereosis,* and others.

Herrick and Alexopoulus (1943) found thiamine in cultures of 22 species of bacteria and actinomycetes.

According to Schmidt and Starkey (1951), 99 species of bacteria and actinomycetes, out of a total of 150 species studied, synthesized heteroauxin. Twenty-two bacterial cultures out of a total of 75 cultures studied, synthesized vitamins on synthetic media.

We have studied 192 bacterial cultures isolated from different soils of the USSR. These cultures were tested for their capacity to synthesize vitamin B_1 and heteroauxin. It can be seen from the table that more than 50% of the bacteria studied synthesize vitamin B_1 and almost 40% synthesize heteroauxin.

Table: *Biotic compounds in bacteria*

Bacteria	***No of cultures tested***	***Number of cultures synthesizing vitamin B_1***	***Number of cultures synthesizing vitamin B_2***
Bac. subtilis	18	10	10
Bac. mesentericus	15	12	5
Bac. sp	13	3	1
Ps. flourescens	8	8	8

Contd...

Bacteria	*No of cultures tested*	*Number of cultures synthesizing vitamin B_1*	*Number of cultures synthesizing vitamin B_2*
Ps. denitrificans	12	10	12
Ps. mycolytica	3	3	3
Bact. coli	2	0	0
Bact. proteus	2	0	0
Bact. liquefaciens	8	5	3
Bact. sp.	13	6	8
Rhizobium trifolii	15	10	0
Rhizobium phaseoli	15	12	0
Rhizobium leguminosarum	8	6	8
Az. chroococcum	16	16	10
Az. vinelandii	4	4	2
Mycob. album	12	7	6
Mycob. citreum	10	10	8
Mycob. rubrum	3	0	0
Microc. albus	3	0	0
Microc. flavus	5	1	1
Microc. aureus	6	0	0

Shavlovskii (1954, 1955) studied cultures of Ps. *fluorescens, Ps. aurantiaca, Bact. herbicola and Ps. radiobacter* from the rhizosphere of plants. These bacteria were grown in synthetic media of a defined composition and after 8 days incubation the vitamin contents of the bacterial cells and of the culture filtrate were determined.

Biotic substances were also found in the mycorhiza fungi. According to Schaffstein (1938), mycorhiza fungi of orchids synthesize growth factors required for the normal growth of the host plant.

Biotic compounds are also synthesized by algae in the soil. It was mentioned above that these organisms are widely distributed in soil. Frequently algae grow on the earth's surface, forming a blue-green coating, visible to the naked eye. It is known that algae secrete various organic substances—metabolic products (Goryunova, 1950). Biotic substances are among these products.

Ondratschek (1940) found ascorbic acid, (vitamin C) in the secretions of such algae as *Hormidium borlowi, H. flaccidum, H. nitens* and *H. stoechidium*. The green alga *Chlorella* synthesizes heteroauxin (Lilly and Leonian, 1941).

The majority of soil microorganisms are partially auxoautotrophs, i. e., they require only some biotic substances. For example, *Clostridium butyricum* requires only biotin and synthesizes its other requisites. Some bacterial species require only thiamine or pantothenic acid. According to Burcholder, McWeigh and Mayer (1944) of 163 yeast cultures 87% required biotin, 35% thiamine and pantothenic acid and 12 % required only inositol.

There are many bacteria which can synthesize only a fraction of a vitamin's molecule. For example, some organisms synthesize only part of the thiamine molecule, either thiazole or pyrimidine components of vitamin B_1. Consequently, the former would require pyrimidine and the latter—thiazole. There are many soil microbes which require ß-alanine, desthiobiotin. pimelic acid and some other compound s—components of this or other molecule of a biotic substance (Meisel, 1950; lerumalimskii, 1949; Stephenson, 1951, and others).

Thompson (1942) has shown that bacteria synthesize vitamins in amounts exceeding those required for their metabolism. The excess of vitamins is released into the environment. The enrichment of the substrate with vitamins takes place, not only at the expense of decomposing cells, but also through the secretion of vitamins by living organisms. The possibility in not excluded that some biotic substances are waste products of growing cells.

According to Thompson, about 50% of the synthesized thiamine remains in the cells of bacteria in a bound state. This fraction finds its way into the soil only after the death and decomposition of the cells.

According to West and Wilson (1938, 1939) there are 19.6 μ g thiamine and 0.37 μ g riboflavin per 1 g of dry weight of the root-nodule bacteria of clover, grown on a synthetic medium. *Clostridium (Clostr. butyricum)* synthesizes 0.9 μ g/g riboflavin and *Micro. ochraceus, Micr. citreus, Ps. pyocyanea,* about 10-15 μ g.

Yamagutschi and Usami (1939) found about 1.5 μ g thiamine per g of dryweight of cells in cultures of *Ps. fluorescens, Ps. alba, and Bact. prodigiosum.* In cultures of *Bact. proteus* they found 9-14 μ g thiamine per 1 g dry weight of cells.

Considerable quantities of biotic substances are synthesized by many mycobacterial species. For example, *Mycob. smegmatis* synthesizes about 135 μ g Oof vitamin B_2 per ml of synthetic medium and 36 μ g per g of the dry weight of cells (Mayer and Rodbart, 1946). Forty to eighty μ g per ml of medium. of the active substance mycobactin was found in cultures of *Mycob. phlei* (Francis at al., 1953).

Ostrowsky and others (1954) studied vitamins in various representative microorganisms.

Table: *Vitamin content of bacterial cells (μ g per g of dry weight of cells)*

Microorganisms	B_1	B_2	*Nicotinic acid*	*Biotin*	*Peteroyl-glutamic acid*	*Pantothenic acid*
Thiobact. thioparus	21	31	92	0.77	0.46	0.75
Thobact thiooxydans	23	60	15	0.64	1.89	57.0
Ps. pyocyanea	15	43	240	2.4	1.0	140.0
Ps. chroococcum	96	--	590	--	--	--
Propionibact. pentosaceum	6.4	--	--	--	--	93.0
Clostr. butyricum	9.3	55	250	1.7	0.5	92.0
Bact. proteus	21.0	--	250	3.4	4.2	100.0
Ps. flourescens	26.0	68	210	7.1	1.8	90.0
Bact. prodigiosum	27.0	35	240	4.1	3.2	120.0

The above-given quantitative data are not strictly constant. They may vary, depending on the growth phase and culture conditions of the individual species of the bacteria, fungi or actinomycetes. In some cases old cells contain less vitamins than young cells, whereas in some species the reverse picture is observed: the old cells contain more vitamins than the young cells. For example, some cultures of *Ps. radiobacter* contain more vitamin B_1 after 8 days growth than after 2 days growth.

In some media the microbes synthesize many growth factors, in others only a few or none. The intensity of the formation of vitamins by soil organisms is greatly influenced by the symbiotic microbes. Some of them suppress vitamin synthesis and other stimulate this process. According to Smalii (1954), *Azotobacter (Az. chroococcum)* in pure culture synthesize 173 μ g of heteroauxin (per cell mass in 1 Petri dish, on Ashby agar) and in the presence of the following microorganisms synthesizes this substance in the noted amounts:

- Bact. mycoides, 220 μ g
- Bact. denitrificans, 196 μ g
- Ps. radiobacter, 243 μ g
- Torula rosea, 234 μ g
- Act. Coelicolor, 188 μ g
- Penicill. Nigricans, 149 μ g

The capacity to synthesize auxins or vitamins does not characterize a species Different strains of one and the same species differ markedly from each other. For example, of more than 100 strains of *Az. chroococcum* which we isolated in different soils and places in the USSR, some of them synthesized large quantities of heteroauxins and others synthesized only small quantities, or none at all. Formation of heteroauxin by the different representatives of these cultures in shown in the coleoptile photograms.

Similar data were obtained when studying other bacterial species and not only for heteroauxin but also for biotin, thiamine, riboflavin, and other biotic substances.

Although there is no strict species specificity as far as the synthesis of biotic substances is concerned, nevertheless mass analysis does show group differences in this respect. More strains of *Azotobacter* synthesize vitamins and auxins than bacteria of the genus *Bacterium.*

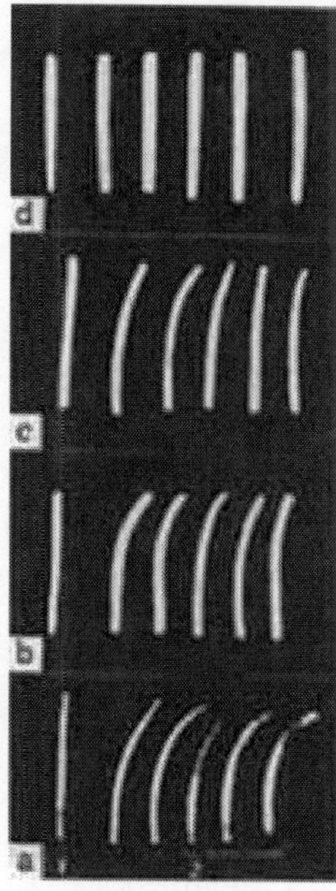

Figure: *The formation of heteroauxin by different cultures of* Azotobacter chroococcum. *The coleoptile curvature after immersion in the culture, expressed in degrees: a) museum strain 54, angle of deviation—32°; b) strain isolated from garden soil in Moscow vicinity, angle of deviation10°; c) strain isolated from the soils of Kara Kum, angle of deviation-8°; d) strain isolated from cultivated podsol soil (Experimental Station Chashnikovo, Moscow Oblast, angle of deviation—0°; 1—control coleoptile; 2—experiment immersed in the bacterial culture.*

Only a few species of root-nodule bacteria are capable of synthesizing heteroauxin and even these are weak forms. We have investigated 12 species of root-nodule bacteria of clover, lucerne, kidney beans, vetch,

Lathyrus vermus, lungwort, peas, Onobrychis, soya, lupine, acacia and astragalus. All these species either did not synthesize heteroauxin at all, or synthesized it in small amounts only.

Out of 60 strains of root-nodule bacteria of lucerne, only 9 strains synthesized this compound in amounts able to give a barely perceptible coleoptile curvature. Fifteen strains of *Rhizobium trifolii,* 8 strains of *Rhizobium leguminosarum* and 15 strains of *Rh. phaseoli* were examined. In all cases the picture was the same.

We have not detected any synthesis of heteroauxins by proactinomycetes which form nodules on the roots of alder tree; actinomycetes and proactinomycetes of the soil do synthesize heteroauxin to a greater or lesser extent.

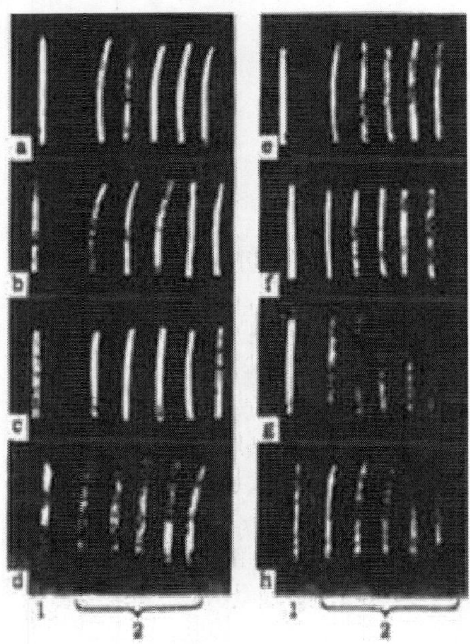

Figure: *The formation of heteroauxins by various species of root-nodule bacteria. The magnitude of the curvature on immersion in cultures of: a) red clover, the angle of curvature-6°; b) soya, angle of curvature- 6°; c) broad beans, angle of curvature- 3°; d) peas, angle of curvature-4°; e) vetch, 4°• angle; f) sweet clover, 2° angle; g) beans, 2° angle, h) proactinomycetes from the nodules of alder tree, 3° angle; 1—control coleoptile; 2—experiment immersed in bacteria.*

According to Starkey (1944), the nicotinic-acid content of plant residues ranges from 2.4 to 85 μ g per gram of dry weight, in the majority of cases it is lower than 30 μ g/g. The same substance in microbial cells amounts to 150-1,920 μ g/g, i. e., approximately 25-60 times more. It should be noted that the studies of biotic substances were carried out on relatively few species of soil microorganisms. The choice of organisms was taken at random, and the studies were confined to a few vitamins only, in the majority of cases to thiamine and riboflavin.

It should be assumed that in reality many and possibly all soil microorganisms synthesize these or other biotic substances which play an essential role in the life and metabolism of lower and higher organisms.

It is obvious that under natural conditions (life in the soil) the microbial metabolism and the synthesis of biotic substances would differ from that under laboratory conditions (on artificial nutrient media). Schmidt and Starkey (1951) have shown that if plant residues which do not contain vitamins are introduced into soil which also does not contain vitamins, the latter appear and accumulate in greater or lesser amounts due to the decomposition of the residues by microbes. The increase in the riboflavin content of the soil is concomitant with the intensification of microbial metabolism. The more plant residues introduced into the soil, the more intense the microbial growth and the formation of riboflavin.

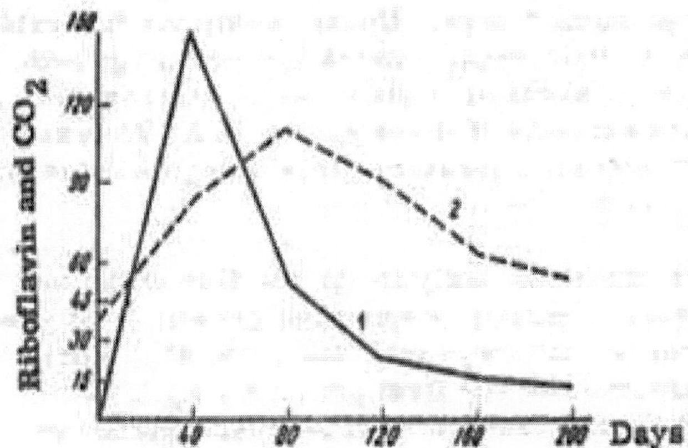

Figure 3: *The formation of riboflavin in the soil as a product of the metabolism of microorganisms, the activity of which is determined by the evolution of CO_2 in mg per 100 g of soil:*

1—riboflavin, in μ g/100 g; 2—CO_2 in mg/100 g.

Table: *The formation of riboflavin in the soil during the decomposition of oat straw (μg per 100 g of soil)*

Accumulation of riboflavin in days:	***0 days***	***1 day***	***3days***	***4 days***	***7 days***	***56 days***
1.25 grams of straw applied	11	19	26	27	26	13
2.5 grams of straw applied	20	22	60	55	38	19

Similar results are obtained if glucose or saccharose are introduced into the soil instead of straw; the bacteria inoculated into the soil lacking the vitamins begin to grow at the expense of the sugars; and riboflavin, biotin, heteroauxins, etc accumulate in the soil.

According to Meisel's calculations (1950), about 400 g of vitamin B_1, 300 g of vitamin B_6 and 1 kg of nicotinic acid are synthesized by microbes in the surface layer of one hectare of the fertile soils of the southern regions, during one season (9 months). Biotic substances are preserved in the soil for varying periods of time. A pure preparation of a vitamin introduced into the soil can be detected for several days. According to Schmidt and Starkey (1951), riboflavin and pantothenic acid persist in the soil from 3 to 20 days or longer.

Table: *The preservation of riboflavin and pantothenic acid in soil (μg/100 g soil)*

itamin	*Amount introduced into the soil*	*Soil*	*Present after 0 days*	*Present after 1 day*	*Present after 2 days*	*Present after 3 days*	*Present after 6 days*	*Present after 21 days*
›oflavin								
	40	Sterile	38	33	--	40	34	34
	40	Nonsterile	36	36	--	43	16	12
	80	Sterile	69	68	--	81	67	64
	80	Nonsterile	65	68	--	81	49	13
ıtothenic ıd								
	50	Sterile	34	34	35	35		
	50	Nonsterile	32	34	10	10		
	100	Sterile	72	77	80	73		
	100	Nonsterile	68	64	18	10		

Riboflavin persists in soil longer than pantothenic acid. Both compounds last longer in sterile than in contaminated soil, since biotic substances, like all other compounds, are subjected to microbial decomposition.

If the microbial metabolism is artificially arrested, vitamins introduced into contaminated soil persist for the same periods as in sterile soil. It was found that biotic compounds (vitamins and heteroauxin) persist in samples of dry soil taken from cultivated and fertilized fields, from 3 to 4 months to 4 years depending on the kind of soil and its properties and also on the properties of the vitamins themselves (Stewart and Anderson, 1942).

Vitamins and other biotic substances entering the soil by one or another route are decomposed and synthesized de novo by microorganisms, Some vitamins disappear others appear. There is a continuous turnover of these substances in the soil. Biotic substances can be found in the soil during the entire vegetative period an long as the microbes live, reproduce and exhibit metabolic activity. The amount of the biotic substances is determined by the rate of their synthesis and introduction into the soil and also by the rate of their destruction, or their stability,

The Effect of Biotic Substances on Plants

It was noted above that green plants synthesize for themselves the necessary biotic substances or phytohormones. Under conditions favorable for their growth this synthesis meets all their requirements for normal growth.

In certain, not infrequent, circumstances, apparently under some unfavorable conditions, the plant synthesizes inadequate amounts of these substances. Then specific avitaminoses develop which are expressed to a greater or lesser degree in the form of certain physiological disturbances and diseases.

Different plants react variously to the addition to the substrate of growth factors and vitamins. Some respond by enhanced growth or by changes in the course of biochemical processes, others react weakly and still others do not react at all. This permits us to assume that the first produce only minimal amounts of the active substances which are insufficient for their normal metabolism, the second synthesize them quite actively but in amounts still insufficient to satisfy all their needs, and the third synthesize them in adequate quantities.

Investigations show that even the last group of plants by no means always synthesize adequate amounts of biotic substances. The vitamin content of plants varies within a wide range, depending on external conditions of growth. It varies according to the soil and climate conditions (Murry, 1948; Rakitin, 1953). Fertilizers have a great effect on the quantity of vitamins present in plants.

In all cases of avitaminosis the vitamins from the substrate are absorbed by the plant. Even under normal conditions of growth, plants utilize ready-made biotic substances, if available.

The utilization of vitamins, auxins and other compounds from the soil has been confirmed in many experiments. Many plants and biotic substances were studied under laboratory and field conditions, in sterile and nonsterile experiments. The plants' requirements for vitamins and auxins has been thoroughly studied in experiments with isolated organs and tissues, and especially with isolated roots. It is known that excised roots of many plants will not grow in synthetic media in the absence of biotic substances and a carbon source. If a root 2-3 mm long is excised from a plant which grew under sterile conditions and placed in a synthetic artificial medium (Bonner's medium or other) it will grow in length to reach considerable dimensions and will form lateral roots, etc, only if the necessary biotic substances are present in the medium. In the absence of the latter, or if their concentration is insufficient. the roots will not grow at all or the growth will be weak.

Investigations show that roots of different plants demand different growth factors. For example, roots of flax require vitamin B_1, roots of peas, horse-radish, lucerne, clover and cotton require vitamins B_1 and B_6; roots of tomatoes thorn apple, and sunflower require vitamins B_1, B_6 and pantothenic acid (Bonner et al., 1937; Robbins and Bartley, 1922-1938; Robbins and Schmidt, 1939, 1945).

Excised roots of many plants, growing on synthetic media, synthesize all the required growth factors. Some of them synthesize them in amounts sufficient for their normal growth, others form too little. The first grow well in artificial media, the latter require the addition of the missing factors (Bonner, 1942). Bonner and Bonner (1948) give the following data on the vitamin requirements of isolated roots.

Table : *Vitamin requirements of isolated roots of various plants (according to Bonner T. and Bonner H., 1948)*

Plants	***Vitamin*** B_1 ***requirement***	***Nicotinic acid requirment***	***Vitamin*** B_6 ***requirement***
Linum usitatissimum Boenn	stimulates	-	-
Raphanus sativus L.	+	+	-
Medicago sativa L.	+	+	-
Trifolium repens L.	stimulates	+	-
Gossypium hirsutum L.	+	+	-
Crepis rubra L.	+	+	-
Cosmos sulfureus	+	+	-
Pisum sativum Gov.	+	+	-
Daucus carota L.	+	-	+
Lycopersicum esculentum Mill	+	-	+
Lycopersicum esculentum pimpinellifolium	+	stimulates	+
Dun	+	stimulates	+
Helianthus annuus L.	+	stimulates	+
Acacia melanoxylon R. Br.	+	stimulates	+
Datura stramonium L.	+	+	+

The following data show the effect of vitamins, on the growth of isolated roots. The roots of flax in the presence of vitamin B_1 elongated by 185 mm. and in its absence by 31 mm in one week. The roots of flax are calculated to synthesize vitamin B_1 at a rate of 0.02 μ g per week. Their vitamin B_1 requirement for normal growth is 2 μ g, i.e., 100 times more than they synthesize.

The roots of white clover grow well in a medium containing vitamins B_1 [The correct designation of this vitamin is unclear.] They increase in length with each successive transfer into a fresh nutrient medium. In the first 5 weeks the roots elongate by 84 mm, the increment in the next 5 weeks amounts to 109 mm, in the third five-week period the increment amounts to 129 mm, in the fourth five-week period—136 mm, and in the following 5 weeks 151 mm. The increment becomes uniform upon subsequent transfer amounting to about 22 mm per week.

The roots of sunflower in the absence of the vitamin complex or in the presence of only one of the vitamins PP or B_6 cease to grow after 7 consecutive resowings. Roots which were supplied with all three vitamins, PP, B_1 and B_6 grew well for along period of time allowing for many transfers into fresh media. In the first five weeks the increment was 74 mm, in the next 5 weeks it amounted to 96 mm, in a further 5 weeks—120 mm, and in the following fortnight—150 mm.

The roots of plants belonging to diverse varieties of one and the same species react differently to vitamins. For example, one variety of tomatoes requires vitamin B_6 and does not respond to vitamin PP and on the contrary another variety requires vitamin PP and does not react to vitamin B_6 (Bonner and Bonner, 1948).

Ovcharov (1955) introduced vitamin PP into a medium in which he grew cotton plants whose leaves were cut off. He observed enhanced formation of new roots on the old roots.

Went, Bonner and Warner (1938) had shown that thiamine stimulates the growth of roots of peas, lemons and camellia. The results were more markedly pronounced when a mixture of thiamine and heteroauxin was employed. Positive results were also obtained in these cases with a mixture of vitamin B_1 and indoleacetic acid (Grebenskii and Kaplan, 1948) and also with vitamin K, biotin and pantothenic acid with biotin (Scheurmann, 1952).

Psarev and Veselovskaya (1947) noted the stimulating effect of thiamine on the formation and growth of wheat roots. In some cases roots of certain plants required only parts of the vitamin molecule, for example only thiazole or pyrimidine (components of vitamin B_1).

Some roots require unknown biotic compounds and cannot, therefore, be grown in vitro.

Robbins (1951) in his review, brings a list of plant species the roots of which can grow on nutrient media. There are 22 such species. Roots of 27 species could not be grown in isolation despite the addition of various vitamins, auxins, amino acids and other biotic substances.

It should be noted that even the roots which can grow in vitro do not grow in the same manner as when attached to the plant. They grow in length and branch, but do not get thicker or if they do thicken, then only very slightly. The activity of the cambium is completely or almost completely suppressed. Consequently, no entirely adequate medium has as yet been found for isolated roots.

The requirement for biotic substances is well pronounced in seedlings. The need embryos of some plants develop better and quicker in the presence of certain vitamins added to the substrate. For example, the growth of pea seedlings separated from the cotyledons considerably increases in the presence of thiamine and biotin (Kögl and Haagen-Smit, 1936). Pantothenic and ascorbic acids also act favorably on pea embryos (Bonner T. and Bonner H., 1948).

Plant embryos do not synthesize biotic substances, they utilize the food reserves present in the seeds. Even the green sprouts of many plants in the early growth period synthesize vitamins weakly or not at all (Bonner et al., 1939). Ripe embryos of thorn apple are easily grown on artificial media without vitamins while the nonripe embryos require vitamins PP, B_1, B_6, C and others.

Pantothenic acid also has a favorable effect on lucerne sprouts. Treating pea seeds with vitamin C enhances their growth by 213% as compared to the controls. Sprouts of meadow grass react positively to the addition of vitamins B_1, PP, H and pantothenic acid to the medium. The grape seeds germinate quicker in the presence of an 0.01 % solution of vitamin PP and, moreover, the formation of roots and the growth of aerial parts is more intense (Flerov and Kovalenko, 1947).

The presowing treatment of cotton seeds with vitamins B_1 and PP considerably enhances their germination and the subsequent growth of the sprouts, Seventy-five per cent of the seeds germinated after treatment with the vitamin as compared to 45 % in the control. The length of the sprouts in the former case was on the average 1.35 cm and 1.65 in the latter Zakharyants, Gorbacheva and Zglinskaya, 1950).

An increase in the growth and subsequent yield of bean seeds after treatment with vitamins B_1 and PP was observed. The height of the plants (from the treated seeds) was greater by 18%, the increment of the vegetative mass was greater by 39%, and the yield of seeds was 28% higher then that of the control plants (Dagis, 1954).

Bonner et al. reached the conclusion that the lower the vitamin content of plant leaves, the stronger they react to the addition of these substances. According to them, peas and tomatoes, contain 13-18 μ g of vitamin B_1 per kg of dry leaves and do not react to its addition. Cabbage, cosmos, Japanese camellia and others contain small amounts of vitamins in their leaves and react positively to the addition of these substances. However, this does not hold for all plants. There are species or even varieties of one and the same species which contain a small amount of vitamins in their leaves and react less to their addition than plants with higher vitamin content.

The addition of vitamins has a favorable effect even on mature plants. Tung trees after the addition of 0.5 mg of vitamin B_1 grew in 70 days twice as much as the controls. The application of low concentrations of vitamin B_1 to poppies increased the weight of their bolls as well as the crop in general. Application of vitamin B_1 together with water had a favorable effect on the growth of spinach. The weight increment during the 63 days of the experiment exceeded many times that of the control plants.

Table: *The growth rate of lemon seedlings under the influence of vitamin PP (according to Kocherzhenko and Snegirev, 1946)*

Treatment	*Average height of plants on 15/ VIII*	*Average height of plants on 15/ IX*	*Average height of plants on 15/ X*	*Average height of plants on 15/ XI*
Control	27.2	29.2	35.5	38.3
Vitamin PP	23.8	34.7	44.5	47.7

Vitamins play a considerable role in the development of orchids. These plants, as already mentioned, grow badly, or do not grow at all without the micorhizal fungi. It was found that the seeds of orchids contain only small amounts of vitamin PP, which are not sufficient for normal germination. This shortage is remedied thanks to the mycorhizal fungi. Treating the seeds with vitamin PP secures their normal germination in the absence of the fungi. The dry weight increment was 3 times higher than that in control plants. It was shown that orchids of the group Vanda grow well in the presence of substances obtained from the mycorhizal fungi. These substances resemble, in their action, bios II (according to Kelly, 1952). According to Noggle and Wynd (1943), some orchids grow well in the presence of nicotinic acid. Henrikson (1951) noted the positive effect of thiamine, vitamin B_6 and nicotinic acid on the germination and subsequent growth of *Thunia marschaliana* Rchb. f..

***Table:** The effect of vitamins on the growth of the orchid Thunia marschaliana Rchb. f. (according to Henrikson, 1951)*

Vitamins	*Height of plants, mm*	*Number of leaves*	*Length of roots, mm*	*Dry weight, mg*
Control	48.0	5	60.5	30.8
B_1	83.5	6	92.5	59.1
B_6	48.0	6	58.5	59.1
C	55.5	6	55.5	30.1
PP	101.0	8	176.5	86.6

Rakitin and Ovcharov (1948) employed vitamin PP and adenine for increasing the growth of cotton plants in their early growth stages. Thereby, not only growth, but also fruiting was increased. The number of bolls was increased, and the cotton yield (raw material) was considerably higher than that of the controls. Similar data were obtained by Zakhar'yants, Gorbacheva and Zglinskaya (1950). They sprayed cotton plants with solutions of vitamin PP and thiamine, The cotton crop increased by 34.1% as compared to the control.

The fat-soluble vitamins, A, E, K, in contrast to vitamins of the B-group suppress the growth and diminish the crop of plants. Carotene suppresses the growth of safranin which is itself rich in carotene. Vitamin K suppresses the growth of fungi, some bacteria and the roots of higher plants. Vitamin PP antagonizes the action of vitamin K. Vitamin E, according to Schopfer (1950), arrests the growth of certain plants. The height of plants in the control was 35.75 cm and in the presence of vitamin E—6.14 cm; the number of flowers in the former was 80.4 and in the latter, 10.

Experiments with vitamins under sterile conditions are worthy of mention. McBorney, Bollen and Williams (1935) tested the action of pantothenic acid on the growth of lucerne under sterile conditions in sand cultures, in a medium which did not contain nitrogen. Pantothenic acid was added in high concentrations. Plants under these conditions grew in the presence of pantothenic acid (in high concentrations of pantothenic acid) much better and the yield was higher. The effect of certain organic substances including some containing vitamins, on the growth of cereal and leguminous plants, in the presence and absence of microorganisms. The experiments showed that in the presence of microorganisms organic substances rich in vitamins have a favorable effect on the growth of plants.

Shavlovskii (1954) tested the effect of pantothenic acid, vitamin B_1, nicotinic acid and vitamin B_6 on the growth of lucerne. The latter was grown on agar medium under sterile conditions for 30 days.

Table: *The effect of vitamins on the growth of lucerne(vitamin concentration in the medium = 0. 1 μ g/ml)*

Vitamins	*Dry-mass weight of 20 plants in mg: Tops*	*Dry-mass weight of 20 plants in mg: Roots*	*Dry-mass weight of 20 plants in mg: Total*
Control (without vitamins)	38.6	8.0	46.6
Pantothenic acid	36.4	11.6	48.0
Vitamin mixture	37.2	12.2	49.4

Analogous experiments were carried out by Shavlovskii with buckwheat. The plants were grown in sand wetted with the nutrient solution of Hellrigel. containing 1 μ g of the vitamin per ml. Other containers were supplemented with yeast extract and vitaminless casein hydrolysate. In one series of experiments the bacterial culture of *Ps. aurantiaca*—vitamin producers were introduced. Plants were grown for 2 days and then analyzed.

Table: *The effect of biotic substances on the growth of buckwheat*

Biotic compound	*Dry mass weight of 10 plants in mg: Cotyledons*	*Dry mass weight of 10 plants in mg: Stems*	*Dry mass weight of 10 plants in mg: Roots*	*Dry mass weight of 10 plants in mg: Whole plant*
Control without vitamins	73.0	64.0	32.0	168.0
Bacteria *Ps. aurantiaca*	76.0	64.0	41.0	181.0
Vitamin B1	82.0	63.0	39.0	184.0
Vitamin B12	79.0	64.0	32.5	175.5
Vitamin mixture	80.0	65.0	36.0	181.0
Yeast extract 0.01%	80.0	66.0	40.0	186.5
Yeast extract 0.1%	90.0	65.0	40.0	195.0
Casein hydrolyste 0.1%	80.0	66.0	43.0	189.0

It can be seen from the given data that the substances tested, markedly increase the increment of the roots and aerial parts of the plants. The biological role of vitamins has been little studied, but, according to the available data, it is important. It is well known that many of them are components of various enzymatic systems. The so-called coenzymes which enter into chemical interaction with the substrate include many vitamins, It has been found experimentally that vitamin B_1 in a compound together with phosphoric acid is the coenzyme of carboxylase-cocarboxylase.

Carboxylase is an enzyme participating in transformations of carbohydrates. It is widely distributed in plants, animals and microbes. Without it the various transformations of carbohydrate compounds, including pyruvic acid, are not feasible. The latter is the key intermediate in the metabolism of living cells, which links the metabolism of carbohydrates, proteins and fats. In the absence or shortage of vitamin B_1 the synthesis of cocarboxylase is slowed down or arrested and, consequently, carbohydrate metabolism is slowed The latter is frequently arrested at the stage of pyruvic said, which leads to the accumulation of pyruvic acid in the cell and the complete cessation of metabolism.

Vitamin B_1 participates not only in decarboxylation of pyruvic acid but also in the reverse reaction—the fixation of CO_2 in pyruvic acid. The role of vitamins in the fixation of CO_2, as the investigations of recent years have shown, is very great.

Vitamins also play a considerable part in the formation and transformation of proteins. It has been shown that vitamins B_2, B_6, B_{12}, PP and H participate in the formation of amino acids and their transaminations. The shortage of vitamin B_6 leads to a decrease in the formation of amino acids from organic acids and ammonia. Vitamin B_e takes part in the formation of amino acids from organic acids and ammonia. Transamination, i.e., transfer of an amino group (NH_2) from one acid to another, takes place in the presence of vitamin B_6. In fat synthesis from sugars, vitamins B_1, B_2, PP and pantothenic acid participate, and the transformation proteins into fats also requires vitamin B_6. Vitamins play an immense role in respiration. It was shown that enzymes participating in respiration consist of proteins and a coenzyme.

The latter consists of vitamin B_2 and phosphoric acid. Vitamin B_2 in enzymatic systems plays a role in oxidation-reduction processes. Folio acid is of great importance in respiration. The germination of seeds and the respiration of sprouts increases under the action of this acid (Stephenson, 1951, Schopfer, 1943, Zeding, 1955). Growth stimulators—auxins and heteroauxins—have a positive effect on the colloidal and chemical properties of protoplasm. According to some authors, they participate in the general metabolism of the cell as separate components. By increasing the metabolism they influence the growth of the cells of the aerial parts and especially of the roots. Under the influence of heteroauxin the influx of plastic substances increases which leads to the formation of now roots in greater quantities. During the rooting of grafts, hydrolysis of starch and fats increases in the cells of the latter. The activity of peroxidase is also increased and the tissues are better hydrated (Maksimov, 1940, Turetskaya, 1955, Zeding, 1955, and others). The action of these substances does not affect the turgor of cells only, as previously assumed, but affects the general metabolism of the plant. In this respect they resemble other biotic substances. Data exist which show that heteroauxin stimulates the formation of auxins, (Zeding, 1955).

Vitamins have a favorable effect on the fertilization of plants. It was found that the sexual organs are rich in vitamins especially the pollen. For example, the pollen of the pea tree contains 2,300 mg of carotene and that of sunflower, 1,460 mg per kg. Pollen of some plants do not contain large amounts of carotene. Vitamins decompose under the action of light and pollen decolorizes and loses its activity. Processing of such pollens with carotene increases its capacity to germinate. Thus, according to Lebedev (1952), without the addition of carotene the percentage of germinated hemp pollen was 39%, the length of the pollen tubes was on the average 100 μ; in the presence of carotene the percentage of germinated pollen was 53.5 and the length of the pollen tubes 312 μ .

The lower the vitamin content, the sharper the reaction to the addition of carotene. Pollens rich in this vitamin stimulate the germination of pollen which contains small amounts of the provitamin if they are left to germinate together. Other vitamins (C_1, B_6, B_1, B_2, PP) also have an effect on the germination of pollen. Pollens of different species and also of different varieties of the same species do not give the same reaction to the addition of vitamins. For example, pollens of one variety of tobacco require 0.0002 mg of vitamin B_1, and pollens of another variety of the same plant require 0.005 mg per liter of the solution. Thirty one per cent of pine pollens germinated in a medium vitamin PP and in the presence of this vitamin 54% of the pollens germinated. About 10-12% of the pollen grains of one variety germinate in the presence of vitamin B_1 and in another variety—52% germinate.

The Assimilation of Biotic and Other Substances by Plants

The problem of plant absorption of biotic substances-vitamins. auxins, and other organic substances, has interested investigators for a long time.

Assimilation of Vitamins

Many investigators have studied the uptake of vitamins and auxins through the root system or leaf surface. Carpenter (1943) introduced riboflavin by spraying the crowns of decapitated plants such as tomatoes, tobacco, fuchsia and carrots, which were subsequently kept in a dark room. The analysis of their sap showed the presence of riboflavin in much higher concentrations than that in the control plants, sprayed only with water. Plants sprayed with a thiamine solution contained thiamine in higher concentrations than the control plants.

Table: *Thiamine concentration in leaves of plants after its artificial application, in mg/kg of dry weight (according to Bonner at al., 1939)*

Plants	Treated Plants	Control Plants
Brassica alba Schmalh.	15.8	6.0
Brassica nigra Koch	6.4	3.9
Agrostia tenuia L.	8.6	5.8
Poa trivalis L.	7.2	4.4
Cosmos Gay	6.0	5.0

Radioactive vitamins are being used in recent years for the detection of their uptake by plants. The plants were grown under sterile conditions on an agar medium of Knopp, supplemented with the mentioned radioactive vitamin at a concentration of 0. 1 y/ ml. The activity of this medium was 4,400 counts / minute (cpm) per 1 ml. The amount of vitamin in the tissues of the plant was determined after various periods of time. The number of cpm showed the amount of absorbed vitamin.

Table: *Absorption of radioactive vitamin B_1 by plants, from sterile agar substrate (according to Shavlovskii, 1954)*

Plant organs	Buckwheat: cpm per plant after 11 days	Buckwheat: cpm per plant after 38 days	Peas: cpm per planlt after 11 days	Peas: cpm per plant after 38 days	Corn: cpm per plant after 11 days
Leaves	3,458	4,920	1,110	1,326	6,993
Stems	1,655	13,288	1,486	2,912	--
Roots	505	1,601	455	3,360	783

These results show that the radioactive (S^{35}) vitamin B_1 enters the plant is the roots, is initially concentrated in the leaves, and then gathers in the stem and in the roots. Apparently, the increase of the vitamin concentration in the roots and stems in the late growth phase of the plant may be explained by the fact that the vitamin is absorbed in the early phase of the growth when the vitamin synthesis by the shoot is inadequate. It has been shown that plants can obtain vitamins from microorganisms. This was demonstrated by the use of radioactive isotopes. If bacteria or yeasts were saturated with an amino acid or vitamin labeled with phosphorus P^{32} or sulfur S^{35} and inoculated into a medium where plants were being grown, the radioactive substances were soon detected in the tissues of plants. Shavlovskii (1954) grew buckwheat in sand with a culture of *Ps. aurantiaca* or yeasts *(Torlopsis utilis, T. latvica* and *Rhodotorula rubra).*

All the cultures were previously saturated with radioactive vitamin B_1 (S^{35}). After 11 days the plants were analyzed for the S^{35} content of their tissues. Simultaneously the excretion of the radioactive vitamin by the bacteria was measured. The analyses showed that buckwheat takes up the vitamin excreted by the microbes in detectable amounts. The greatest amount went to, the plant (8.5 % of the total activity) from *Ps. aurantiaca.*

Table: *Transfer of vitamin B1 from microbes to buckwheat (according to Shavlovskii, 1954)*

Part of plants	***S^{35} (of vitamin B_1) cpm/plant Ps. aurantiaca 8,100 cpm***	***S^{35} (of vitamin B_1) cpm/plant T. latvica 45,000 cpm***	***S^{35} (of vitamin B_1) cpm/plant Rh. rubra 9,100 cpm***	***S^{35} (of vitamin B_1) cpm/plant T. utilis 10,230 cpm***
Cotyledons	294	252	144	150
Stems	217	161	96	114
Roots	180	109	77	77
Total per 1 plant	691	522	317	341
Given up by microbes to plant, in percent	8.5	1.2	4.0	3.3

Assimilation of Amino Acids

The capacity of roots to absorb amino acids has been proven experimentally. Petrov (1912) gives data on the absorption of asparagine by plants (corn). Shulov (1913), Pryanishnikov (1952) and Byalosuknya (1917) confirmed these data. According to them, asparagine is a good source of nutrition for peas, corn, cabbage, mustard, and other plants.

Virtanen and Lincol ascribe a stimulating role to the amino acids. According to them, small concentrations of alanine and phenylethylamine (decarboxylation product of phenylalanine) strongly affect the growth of peas, in the same way as the heteroauxin. The plant parts change markedly, becoming stronger and greener. In order to demonstrate the uptake of amino acids by plants, Shavlovskii (1955) studied the absorption of, radioactive methionine S^{35} by buckwheat, corn, and peas.

This culture excretes more vitamin than the others. Yeasts excreted various amounts of the vitamin depending on the species. The data given below should be considered as approximate and possibly lower than in reality. This amino acid was added to the medium of Knopp in which these plants were grown. After 11 days of growth the plant tissues were examined for the presence of radioactive sulfur. The results are given.

Table: *The absorption of radioactive methionine S^{35} by plants (cpm per plant) (according to Shavlovskii, 1955)*

Plants	*Specific activity: Leaf*	*Specific activity: Stem*	*Specific activity: Root*	*Activity, (dry weight) of: Leaf*	*Activity, (dry weight) of: Stem*	*Activity, (dry weight) of: Root*
Buckwheat	56	81	625	389	759	1,389
Corn	65	--	124	5,814	--	3,452
Peas	46	31	628	828	992	7,530

As can be seen from the table, the absorption of this amino acid by plants is quite vigorous. The highest concentration of the amino acid in in the roots. The intensity of absorption varies according to the composition of the medium and external conditions.

The absorption of methionine is more rapid and more pronounced in the presence of vitamins B_1 and B_6.

Ratner and Dobrokhotova (1956) have shown the activating effect of thiamine pyridoxine and pantothenic acid on the synthesis of glutamic acid and alanine in the roots of sunflower.

The distribution of the radioactive sulfur of methionine is different from that of radioactive inorganic sulfur. In the former came sulfur is concentrated in the roots and in the latter, in the aerial parts (Thomas and Hendricks, 1950, and others). This gives grounds for the assumption that methionine as well as other amino acids are assimilated by plants without any change in their molecules and are being used by them for protein synthesis. Shavlovskii extracted protein from the roots of peas which had grown in the presence of radioactive methionine and showed that they contained the major part of the methionine absorbed by the roots. One gram of fresh roots gave 23,171 cpm, and the proteins isolated from them—13,220 cpm.

Plants absorb the amino acids synthesized and excreted by microbes. This was shown by Shavlovskii who used methionine containing radioactive sulfur S^{35}. As in experiments with vitamins, Shavlovskii grew bacteria *Ps. aurantiaca* and yeasts *Sacchar. cerevisiae,* in a medium containing radioactive sulfur S^{35} in the form of $(Na_2 S^{35}0_4)$. The bacteria were then autolyzed and the autolysates containing radioactive (S^{35}) amino acids were added to the medium where plants were grown, Radioactive sulfur was then found in the plant tissues; the roots showed more radioactivity than the aerial parts. In the presence of autolysate of *Ps. aurantiaca* the activity in the cotyledon tissues was 137 cpm, in the stems—356 cpm, and in the roots—720 cpm; in the presence of yeast autolysate the corresponding figures were: 43 cpm, 199 and 569 cpm.

In his subsequent experiments Shavlovskii grew buckwheat in a medium with living cultures of *Ps. aurantiaca* previously grown in a medium containing radioactive sulfur. Seeds of the buckwheat were infected with these cells before sowing (700 million cells per seed, with a total activity of 125,000 cpm).

The following activity was found on the second day of growth in the plant: cotyledons—228, stems—181, roots—132 cpms. Consequently, on the second day the plants had taken up about 0.4 % of the bacterial radioactivity. A direct relationship between the amount of bacteria and the absorption of radioactivity was noted. Upon introduction of 3.45 billion cells per seed, about 1% of the total radioactivity of the cells was detected in the plants (after 7 days growth).

The capacity of microorganisms to transfer their metabolic products to the plants was shown in a paper by Akhromeiko and Shestakova (1954). These authors, in their studies, employed radioactive phosphorus P^{32}. They grew *Az. chroococum, Ps. fluorescens* and yeasts (isolated from soil) in media containing P^{32} as a source of phosphorus nutrition. The cultures of microorganisms thus grown were carefully washed with water and inoculated into the sand in which saplings of oak and ash trees were grown.

Experiments had shown that radioactive phosphorus is taken up by the plants in considerable amounts; it is emitted at a different rate and in different amounts by the various microbial species. The greatest effect is obtained in experiments with yeasts. About 43 % of the radioactive phosphorus from yeast was taken up by plants in the first days of their growth. Oak saplings absorbed more radioactivity than ash-tree saplings.

These experiments show that biotic substances formed by microbes, in addition to the amino acids and other metabolites, are excreted into the substrate, and from the substrate are absorbed by plant roots.

Assimilation of antibiotics

Higher plants absorb not only vitamins, auxins and amino acids but also many other organic compounds present in the soil and formed by microorganisms. Of all the metabolites which can serve as indicators of absorption by plants, antibiotics, in our opinion, are the most outstanding. Antibiotics are very specific. They are not present in plant tissues and are not formed by them. They are easily detected and differentiated from other organic substances including phytocides. In our experiments we have employed antibiotics in their native state as well as in the form of chemically pure preparations. Antibiotics produced by different representatives of soil microorganisms were used: penicillin (mold product), streptomycin, globisporin, aureomycin, terramycin, and others (products of actinomycetal metabolism), subtilin, gramicidin, and others (of bacterial origin).

The crude as well as the chemically pure preparations were added, in various concentrations, to substrates where plants were grown. Experiments had shown that antibiotics were taken up by roots rapidly and in considerable amounts and were more or less uniformly distributed in all plant tissues and organs. In a manner similar to that of biotic substances and amino acids the antibiotics concentrate in the root system more than in other parts. From there they enter the aerial parts. Antibiotics are absorbed by plants directly from the soil where they are formed by microorganisms. The latter, as it will be shown below, developing in soil under certain conditions, produce and accumulate considerable amounts of these active metabolites, These naturally formed substances enter the plant tissues via the root system in the same way as the chemically pure antibiotics.

The antibiotics are known to be rather complex organic substances of high molecular weight. For example, streptomycin consists of three basic groups: N-methyl, S-methyl and also carbonyl group. Its formula is $C_{21}H_{39}O_{12}N_7$. Its molecular weight is more than 500. No less complex are aureomycin, terramycin, penicillin and other antibiotics.

If such complex organic compounds as antibiotics, amino acids and vitamins are absorbed it may be assumed that many other carbon and nitrogen compounds present in the soil are also absorbed by plants. A voluminous work was performed in the laboratory of the famous French botanist Bonne to determine the absorption of organic compounds by plants. Laurent J. and Laurent J. (1903). studied assimilation of many organic compounds by plants. According to their data, peas, lentils. corn, rice, and wheat utilize glucose, saccharose, glycerol, dextrin, starch and potassium humate. Lefevre (1905, 1906) noted the capacity of plants to assimilate amino acids and other nitrogenous compounds. Ravin (1913), experimenting with horse radish came to the conclusion that plants can assimilate the organic acids—succinic, citric, malic, tartaric and oxalic.

According to Pryanishnikov, 1952; Shulov, 1913; Byalosuknya, 1917; and others, peas, corn, cabbage, buckwheat and other plants assimilate well glucose, saccharose and lactose. Especially good growth of plants was obtained in the presence of levulose. Numerous experimental data as well as agrobiological observations exist which confirm the possibility of heterotrophic nutrition of plants.

Effect of Bacteria on the Assimilation of Nutrients by Plants

The assimilation of mineral and organic nutrients by plants proceeds at various intensities and depends not only on the composition and properties of the compounds in question but also on the quantitative and qualitative composition of the microflora around the root system.

Studies show that under conditions of sterile growth of plants these substances are not absorbed to the same extent as in the presence of microorganisms. We have grown wheat, peas and corn in a sterile nutrient solution and followed the absorption of penicillin, streptomycin, aureomycin and other antibiotics in the presence and absence of various bacterial species. Bacterial cultures were chosen which did not inactivate or decompose the above-listed antibiotics and were at the same time resistant to them for the inoculation of the nutrient solution. We have tested and chosen more than 10 cultures belonging to different species: 3 cultures of root-nodule bacteria *(Rh. trifolii, Rh. meliloti and Rh. phaseoli),* 2 strains of azotobacter *(Az. chroococcum),* 4 strains of *Ps. fluorescens*, isolated from the rhizosphere of different plants, and 2 strains of the genus *Bacterium (Bact. denitrificans* and *Bact.* sp.) also isolated from the rhizosphere of wheat a nd peas. The effect of the bacteria was determined by the rate of disappearance of the antibiotics from the solution and their uptake by the plants. The presence of antibiotics in the plants was determined by the conventional microbiological tests. The indicator microbes were cultures of the sporiferous bacillus *Bact. subtilis* and staphylococcus—*Staph. aureus.*

The analyses have shown that the increased concentration of antibiotics in the plants is accompanied by decreased concentration of antibiotics in the solution. Analogous data were obtained in experiments with plants growing in sand. The uptake of antibiotics was highest in the presence of *Azotobacter* and root-nodule. bacteria and the smallest in the presence of the nonsporiferous bacillus *Bacterium* sp. No. 25. The experiments have shown that under sterile conditions, without bacteria, the uptake of phosphates proceeds at a lower rate than in the presence of bacteria. The greatest effect was obtained in the experiments with buckwheat. in the presence of specially chosen bacteria.

Bacteria markedly increase the rate of phosphorus uptake by plants from granules prepared with radioactive superphosphate. Barley sprouts grown under sterile conditions gave 400—500 cpm, but 1,000-

1,500 cpm, when grown under nonsterile conditions (cpm. per 10 mg of plant dry weight) (Kotelev, 1955). We employed the tracer-atom technique to carry out a number of experiments on the uptake of phosphorous compounds from a substrate in the presence and in the absence of different bacterial species. Plants (barley) were grown in a nutrient substrate (solution, sand and soil), previously sterilized and then inoculated with pure cultures of bacteria. Radioactive phosphorus (P^{32}) was then added to the substrate. After 15-20 days of growth the plants were analyzed. We determined the presence of organic and inorganic phosphorus, as well as sparsely soluble (in water) phospholipides, and phosphoproteins.

In addition, amino acids were determined (by the use of paper chromatography) in the fraction containing organic water-soluble compounds (Krasil'nikov and Kotelev, 1956). Bacteria isolated from the rhizosphere of corn, grown in Moldavia, were used in these experiments. All of them belonged to the nonsporiferous bacteria of the genera *Bacterium* and *Pseudomonas*. One culture (10-A) was isolated from podsol soil of the Moscow Oblast'. Each vessel filled with sand was supplemented with 465 mg P205 (P^{32}). The total activity per vessel was 6,400,000 cpm. Counting was performed after 17 days growth. The radiochromatogram, represents the process of accumulation of organic phosphorus in the tissues of barley grown under sterile conditions, and in the presence of bacteria. It can be seen from the figure that bacteria have a considerable effect on the uptake and accumulation of phosphorous compounds in the tissues of plants. Under the influence of the bacteria a formation of quite different phosphorous substances takes place.

The results of the determination of amino acids are not loss demonstrative. We have determined the amino acids present in extracts of 70% alcohol by paper chromatography. Five hundred-mg samples of the dry mass of plants were extracted for 4 hours at 45° C with 25 ml of alcohol. The extracts were then filtered, the filtrate was evaporated to 1-2 ml and subjected to chromatography.

The radiochromatogram shows the distribution of amino acids in plants grown in the presence of 2 bacterial cultures (lp and 2p). It can be seen from the figure, that the presence of bacteria in the substrate causes formation and accumulation of amino acids of a different nature to those formed in the absence of bacteria. Some amino acids (serine, glycine, alanine, valine, cysteine) cannot be detected, in the control sprouts of the barley grown under sterile conditions, only their traces can be found. These amino acids are present in considerable amounts in plants grown in the presence of bacteria. Different bacterial cultures have a varying effect on the formation and concentration of amino acids in plant tissues. In our experiments cultures 125 and 151-A favored the accumulation of glutamic acid, and bacteria 1p and 2p favored the accumulation of alanine, lysine and asparagine.

The effectivity of the same bacteria to plants growing in soil (heavy loam chernozem), although considerable. was somewhat different to that noticed when the plants were grown in pure sand. The quantitative and qualitative relationship of amino acids was different. Neither lysine nor alanine could be found in the barley sprouts grown in the presence of lp and 2p bacteria, asparagine and glycine were found only in traces.

According to Akhromeiko and Shestakova (1954), microorganisms inhibit the uptake of phosphorus (P^{32}) by woody plants. Saplings of oak and ash tree were grown in sterile and nonsterile soil and the uptake of radioactive phosphorus was determined. The authors have noticed that bacteria of the rhizosphere at first take up the tracer phosphorus and subsequently release it. The difference in composition of the amino acids is well defined on the chromatogram. The addition of microbial metabolic products, or a dead culture, to plants in their early growth stage on sterile medium, causes an increase in the phosphorus and nitrogen content of their exudate. The microbial metabolic products increase not only the uptake of these substances by the roots but also the synthetic capacity of the roots. The increased incorporation of the radioactive phosphorus P^{32} into the lipides and nucleoproteins, as well as the increased content of amide and amine nitrogen in the exudate confirms these assumptions.

Microorganisms and their metabolic products affect the process of nitrogen transformation in the root. In the presence of microorganisms the rate of metabolism of amino acids in the roots increases as well as the process of transformation of the inorganic to organic nitrogen. In the presence of microorganisms the uptake of inorganic and organic compounds—microbial metabolites—is increased. Besides, microorganisms promote the transport of nutrients in the soil. They are carriers of nutrients, supplying the root system with various nutrients.

Numerous papers have stressed the point that mere contact of roots with the soil is not sufficient to secure the nutrient requirements of the plant. Mediators between nutrient sources of the soil and the root system exist in soils. Such mediators are the microbes. Not only fungi but also bacteria promote the transportation of nutrients in the soil. Kotelev (1955) employed tracer techniques in the following manner. He introduced grains of radioactive P^{32} in the form of superphosphate into the soil and followed the diffusion of the phosphorus in the presence and absence of bacteria.

The diffusion of phosphorus in sterile soil was very slow and uptake by roots was either lacking or negligible. The diffusion of phosphorus in the presence of bacteria was much more rapid. The latter are taken up by the roots and aerial parts of the plant and increase the intensity of the biochemical and biological processes in the cells and tissues. They enhance the growth of plants and increase the absorption capacity of the roots the assimilation of absorbed substances and other vitally important functions.

The biologically active substances of the soil not only enhance the growth and increase the yield of plants but also confer on the plant better nutritional qualities. Plants which obtain vitamins and other organic compounds from the soil in adequate amounts, yield crops of higher quality and their seeds are of a higher vitality.

The fact that plants can grow in pure mineral nutrient media in the absence of microorganisms cannot serve as proof of the uselessness of the latter in the nutrition of plants. Plants can indeed be grown in mineral media and yield seeds without the participation of microbes. Is it possible, however. to secure the vitality of such plants in subsequent generations? We have presented our observations on the growth of green algae and duckweeds. Grown under sterile conditions, in mineral media without the addition of composts or metabolic products of bacteria, these plants lose their viability and eventually die out. Plants grown in the same media but supplemented with composts or metabolic products of bacteria, as well as plants grown in the presence of living bacterial cells (nonsterile conditions) have been kept in our laboratory for more than 20 years without any visible decrease of vitality.

The assumptions of some authors concerning the fact that soil contains only small amounts of organic compounds of phosphorus and nitrogen (in the form of vitamins, amino acids and phytin) which cannot, therefore, be considered as nutrients of any great importance, are also groundless. Our knowledge of forms of organic compounds and of the dynamics of their transformations in the complex microbial coenoses, is inadequate. The knowledge we do possess, however, allows us to assume that the processes of synthesis of various organic compounds proceed incessantly in the soil. Owing to such uninterrupted synthesis (be as it may in small amounts) the total production of these compounds may be sufficient to meet the needs of plants.

Bacterial life in the soil is known to be very short; it is counted in hours. Even during their life, the dying cells are subjected to autolysis. At the end of the enzymatic lytic processes, processes of solubilization of their residues by enzymes of other microbes ensue. The process of bacterial cell destruction is rapid, The metabolism of living cells is an endless sequence. The elements absorbed from the substrate are soon excreted. Experiments with radioactive elements have shown that phosphorus (P^{32}), for example, appears in the substrate after a few minutes. The microflora of the root zone is of great importance in plant nutrition. Growing near or on the roots, microorganisms, together with the plants, create a special zone—the rhizosphere. Soil in this zone differs in its physical, chemical and biological properties from the soil outside the root zone. It possesses different conditions, for the absorption and excretion of substances by the roots. The interaction between microorganisms and plants on one hand, and between individual microbial species and their metabolites, on the other hand, is the basis for the different transformations of inorganic and organic compounds. As a result, compounds which serve as nutrients for plants are formed. These are absorbed by the roots.

Substances present in the soil are subjected to a greater or lesser extent of processing before their absorption by the roots. The plants do not absorb those compounds which are characteristic for this or other soil, but metabolic products of the rhizosphere. The rhizosphere microflora prepares various organic and inorganic nutrients for the plants. The role of the rhizosphere microflora reminds one of the digesting organs of animals. Microorganisms in the final account serve the same function in the plant nutrition as the digestive system of animal organisms. The same point of view is held by the American specialist Prof. Clark (1949). He considers that microorganisms living in the rhizosphere perform the same work as do the intestines of animals. Academician Lysenko (1955) is even more definite in this respect. He thinks that the microflora of the root zone acts as the digestive organ of plants. We can agree with such a comparison if we recall the function of the microflora of the intestines. In recent years, the microflora of the intestines has more and more come to be considered as a factor of supplementary nutrition for animals and human beings. Many intestinal bacteria are known to produce substances which enter the organism of the animal and play an essential role in biological processes as biocatalysts.

In many animals, microorganisms of the digestive tract participate directly the digestion of food. For example, the cellulose bacteria in the intestinal tract of ruminants, decompose cellulose into digestible products.

Chapter 6

Field Sampling

Compiling and Presenting Data

When you have finished identifying your samples, compile the data into an Excel file (online data template for modified protocol coming soon). Each team should first organize its own data, and then combine data withother teams. Notes for the class room: This task can be completed as homework and then presented in class with visual aids. Try presenting not as teams but as 'tide levels' in order to form an over all picture of your nearshore.

Presentations should include the following information:

- Introduction to the tide level and area being presented
- Raw data:
  a) Percent coverage and stripe/stem count of the macrofaunal groups
  b) Number of individuals of each taxa level (family, species etc) of the macrofauna
- Comparisons between the findings in the different replicates in the same tide level
- Summaries of what can be found in the area at that tide level
- Extras such as comments on things that you did not expect to find or things that you had difficulty identifying,

Discussion topics to follow the presentations:

- How are the tidal heights different in terms of taxa/species composition?

- Which tidal height has the greater taxa/species richness? Taxa/species evenness?
- Which tidal height has the greater overall diversity?
- Add "Why might this be?" to each of the above points.
- How do your results compare with your initial impressions?

Moving Forward

Delve deeper in to the workings of the near shore. Go beyond simple species lists. Contemplate how your species interact i.e. Outline local food webs, trophic levels, and ecological niches. Consider sampling on a regular bases maybe seasonal or yearly. By accumulating data comparison over time and through various local changes (large storms, construction etc) can be presented.

Physical Properties of Grassland and Cultivated Soils

Soil structure is highly sensitive to human activity and in many parts of the world there is evidence that it is becoming less favourable for agriculture. Since numerical values and indices almost always depend on the measurement procedure, there is no general agreement on criteria to judge soil structural decline under different soil management practices.

Thus, when dealing with the impact of cultivation on soil structure, a large number of methods have been used. The main effect of the diversity of procedures was to minimize the possibility of inter-experiment comparisons.

Some structural characteristics, such as porosity, strength and stability are considered to be most important, because they influence a multitude of processes.

Central to the assessment of soil structural quality and increasingly used is pore space analysis.

Soil pore space can be partitioned into a structural compartment and a textural one. Soil textural porosity represents the pore space due to packing of soil. The structural pore space results from the arrangement of aggregates and depends on factors such as climate, soil organisms and cultivation.

Textural porosity can be calculated from soil bulk-density data determined on small volume samples. This measurement standardization, allows the testing of the consistency of the above-mentioned pore space model and also increases the possibility of inter-experiment comparisons.

One method commonly used for assessing the effects of cultivation on the soil system is to compare grassland and cropland. In the present study changes in soil structure due to tillage practices on moderately coarse to medium-textured soils were investigated. These were assessed relative to adjacent grassland areas.

Material and Methods

The investigated soils are located in Galicia (Northwest Spain). Clods were collected from the surface horizon (0-20 cm) of grassland and cultivated soils. Site characteristics were described in detail by Fernández Rueda (1977).

Aggregates of 2-3 mm diameter were obtained by breaking and seaving the air-dried clods. The assessment of volume and pore space change with associated change in water content was made using Archimedes principle.

The bulk density was determined on small fragments (2-3 mm) as described by Monnier *et al.* (1973). Particle density was determined by the water picnometer method.

The degree of shrinkage, S_p, potentially achieved by the soil aggregates was estimated as proposed by Stengel *et al.*, 1984. Mercury intrusion porosimetry was carried out on dried 2-3 mm aggregate samples, with a Micromeritics porosimeter. The textural pore espace was partitioned into lacune pores and pores of the clay fabric. Water retention curves were also determined on 2-3 mm aggregates in a pressure cell.

Results and Discussion

The results of the present investigation successfully condense those of previous work, namely that the soil in the grassland treatment has better structure.

The most appropriate index to describe structure quality of the studied soils is dry bulk density.

The term resilience is used to designate the ability of a soil to recover its structural form through natural processes, when external forces are removed. The index of potential shrinkage, Sp, derived from the shrinkage curves, was found to be a useful indicator to estimate the potential recovery of soil structure by regenerative processes, i.e. the soil resilience.

Low Sp values can be interpreted as a lack in the properties which promote natural structure regeneration. Some of the investigated horizons demonstrate severe limitations for maintenance of good structure and, in general, a lack of potential for regeneration of soil structure by fragmentation was deduced from the shrinkage curves.

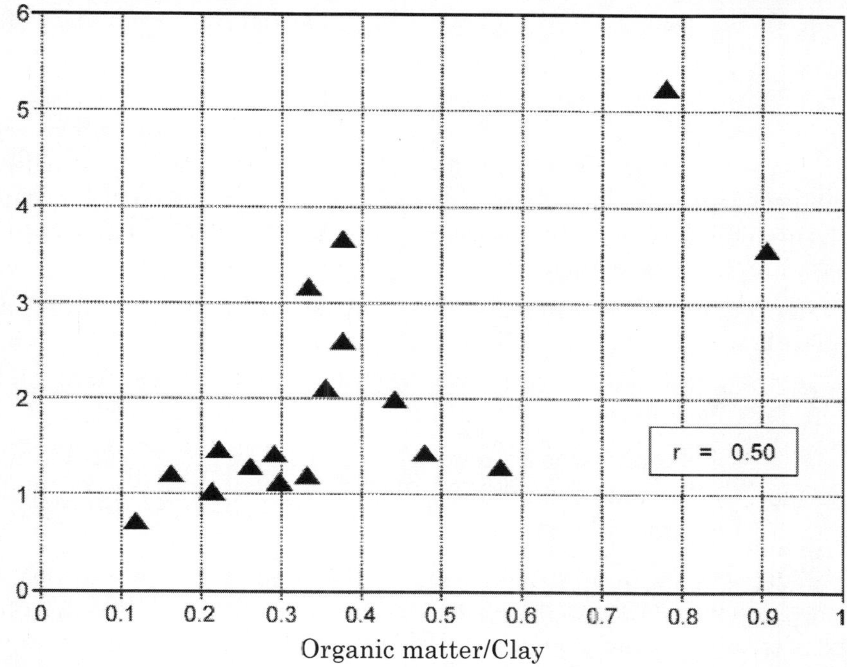

Figure: *Relationship between organic matter content and shrinkage index.*

The organic matter content of the investigated soils varies between 20.3 and 130.4 g/kg. Within this range a significant relationship ($p<0.05$) between Sp index and organic matter content was found. Therefore, the amount of organic matter influences the swelling-shrinking magnitude from water saturation to dry state. This influence appears to be more important at higher water content.

Total porosity measured by Hg intrusion in the range 400-0.006 mm was lower in all cases but lowest in the cultivated soils than in the corresponding uncultivated. Generally, differences were more pronounced where each cultivation practice lasted longer.

Differences between cultivated and uncultivated counterparts could be better appreciated taking into account the lacunar pore range. Thus, the cultivated soils generally showed a decrease of larger pore classes with respect to the corresponding uncultivated soils.

Water retention near saturation was increased with increasing organic mater content, but for potentials of about 1 500 kPa water retention curves determined in samples from adjacent plots tended to converge. This confirms that the organization of the clay fabric is not modified by mechanical tillage, as previously demonstrated by Hg intrusion porosimetry.

This study corroborates previous research, indicating that mechanical tillage leads to a marked increase in dry bulk density. A lack of potential for regeneration of good soil structure by fragmentation was deduced from the shrinkage curves. Mercury intrusion porosimetry showed that lacunar pores prevailed, whose volume increased as organic carbon content increased. Water content near saturation increased also with increasing organic matter content. The possibility of quantifying differences between tillage and no-tillage plots by assessing the specific volume dynamics on soil layers with limited amounts of swelling materials was demonstrated

Soil Properties

Soil surveys of every county in North Dakota have been completed by the Natural Resource Conservation Service (NRCS, formerly the SCS). The county soil survey report provides detailed soils information on any parcel of land and is available from the county NRCS office or the NDSU Department of Soil Science. The soil properties of texture, structure, depth, permeability and chemistry play an important role in irrigation management.

Soil Texture

Soil texture is determined by the size and type of solid particles that make up the soil. Soil particles may be either mineral or organic. In most soils, the largest proportion of particles are mineral and are referred to as "mineral soils." For mineral soils, the texture is based on the relative proportion of the particles under 2 millimeters (mm) or 5/64th of an inch in size.

The largest particles are sand, the smallest are clay, and silt is in between. The soil texture is based on the percentage of sand, silt and clay. Soil texture classes may be modified if greater than 15% of the particles are organic. Soil particles greater than 2 mm in size are not used to determine soil texture. However, when they make up more than 15% of the soil volume, the textural class is modified.

Soil texture can be determined by separating and weighing the sand, silt and clay. For example, if a 100 pound sample of soil was sifted through screens and found to contain 45 pounds of sand, 35 pounds of silt and 20 pounds of clay, then the soil would be composed of 45% sand, 35% silt and 20% clay. This soil has a loam texture. There are 12 basic soil textures Sand, loamy sands and sandy loams are the most common soil textures irrigated in North Dakota.

Soil Structure

Soil structure refers to the grouping of particles of sand, silt, and clay into larger aggregates of various sizes and shapes. The processes of root penetration, wetting and drying cycles, freezing and thawing, and animal activity combined with inorganic and organic cementing agents produce soil structure. Structural aggregates that are resistant to physical stress are important to the maintenance of soil tilth and productivity. Practices such as excessive cultivation or tillage of wet soils disrupt aggregates and accelerate the loss of organic matter, causing decreased aggregate stability.

The movement of air, water, and plant roots through a soil is affected by soil structure. Stable aggregates result in a network of soil pores that allow rapid exchange of air and water with plant roots. Plant growth depends on rapid rates of exchange. Good soil structure can be maintained by practicing beneficial soil management such as crop rotations, organic matter additions, and timely tillage practices. In sandy soils, aggregate stability is often difficult to maintain due to low organic matter, clay content and resistance of sand particles to cementing processes.

Soil Series

Soil is the layer of the earth's surface which has been changed by physical or biological processes. The five soil-forming factors that control the process of change are parent material, climate, topography, biota (plants and animals) and time. Soils are grouped into categories according to their observed properties. The USDA classification system consists of six categories. The highest category (soil order) contains 11 basic soil groups, each with a very broad range of properties. The lowest category (soil series) contains over 12,000 soils, each defining a very narrow range in soil properties.

North Dakota has 264 soil series. A soil series is unique because of a combination of properties such as texture, structure, topographic position (on the side of a hill or in a valley) or depth to the water table. A particular soil series describes locations where these soil conditions are similar.

These locations may be in the same field, section, county, state or even region. Soil delineations on county soil survey maps are based on the soil series. A soil series is generally named after a town near the site that represents the typical properties for that soil. For example, the site with typical properties for the Embden soil series is near Embden, North Dakota. Many soil series do not have a deep, uniform soil profile. Restrictive subsurface layers often interfere with root penetration. In these situations the roots will be concentrated in the upper part of the soil profile. For example, in the Renshaw loam profile, the majority of the plant roots will be in the top 18 inches because of the poor growing environment encountered in the underlying sand and gravel substrata. This type of information is important for irrigation management.

Soil Depth

Soil depth refers to the thickness of the soil materials which provide structural support, nutrients, and water for plants. In North Dakota, soil series that have bedrock between 10 and 20 inches from the surface are described as shallow. Bedrock between 20 and 40 inches is described as moderately deep. Most soil series in North Dakota have bedrock at depths greater than 40 inches and are described as deep. Depth to contrasting textures is given in the soil series descriptions in the county soil survey report.

The depth to a contrasting soil layer of sand and gravel can affect irrigation management decisions. If the depth to this layer is less than 3 feet, the rooting depth and available soil water for plants is decreased. Soils with less available water for plants require more frequent irrigations.

Soil Permeability and Infiltration

A soil's permeability is a measure of the ability of air and water to move through it. Permeability is influenced by the size, shape, and continuity of the pore spaces, which in turn are dependent on the soil bulk density, structure and texture. Most soil series are assigned to a single permeability class based on the most restrictive layer in the upper 5 feet of the soil profile. However, soil series with contrasting textures in the soil profile are assigned to more than one permeability class. In most cases, soils with a slow, very slow, rapid or very rapid permeability classification are considered poor for irrigation.

Table: *Soil Permeability Classes.*

Classification	Infiltration Rate (inches/hour)
Very Slow	Less than 0.06
Slow	0.06 to 0.2
Moderately Slow	0.2 to 0.6

Contd...

Classification	Infiltration Rate (inches/hour)
Moderate	0.6 to 2.0
Moderately Rapid	2.0 to 6.0
Rapid	6.0 to 20.0
Very Rapid	Greater than 20.0

Infiltration is the downward flow of water from the surface through the soil. The *infiltration rate* (sometimes called intake rate) of a soil is a measure of its ability to absorb an amount of rain or irrigation water over a given time period. It is commonly expressed in inches per hour. It is dependent on the permeability of the surface soil, moisture content of the soil and surface conditions such as roughness (tillage and plant residue), slope, and plant cover. Coarse textured soils such as sands and gravel usually have high infiltration rates. The infiltration rates of medium and fine textured soils such as loams, silts, and clays are lower than those of coarse textured soils and more dependant on the stability of the soil aggregates. Water and plant nutrient losses may be greater on coarse textured soils, so the timing and quantity of chemical and water applications is particularly critical on these soils.

Saline and Sodic Soils

Salt affected soils are grouped according to their content of soluble salts and sodium. Saline and sodic soils usually occur in areas where ground water moves upward from a shallow water table close to the soil surface. The water carries salts which accumulate in the soil as the water is evaporated from the soil surface or transpired through the plants to the atmosphere. In general, these soils are not recommended for irrigation. Saline and sodic soils may be of natural or man-made origins. One of the man-made processes is related to irrigation. Under certain combinations of irrigation water quality and soils, salts and/or sodium may accumulate in the root zone and have an adverse effect on plant growth.

Under some conditions, *sodium* can be controlled in the upper part of the soil through the use of calcium amendments. The replacement of sodium by calcium improves the structure of the soil. Calcium soil amendments can be helpful in situations where land with a majority of unaffected irrigable soils contains pockets (inclusions) of sodium affected soils. Under irrigation, calcium soil amendments will help where surface crusting has become a problem. Special irrigation management practices may be required on these soils.

Table: *Soil chemistry measurements used to classify saline, sodic and saline-sodic soils.*

	Electrical Conductivity* (mmhos/cm)	***pH***	***Sodium Adsorption Ratio* (SAR)***
Saline soil	greater than 4	less than 8.5	less than 13
Sodic soil	less than 4	8.5 to 10	greater than 13
Saline-Sodic soil	greater than 4	less than 8.5	greater than 13

**Measured from a saturated soil extract*

Salt concentrations can be managed by leaching or controlling the water table elevation. Leaching is accomplished by applying more water than the soil will hold within the root zone. Large rainfall events, applying additional irrigation water or both will carry some of the salts below the root zone.

Water table control can be accomplished by planting a deep rooted crop, such as alfalfa, or installing subsurface drainage. Deep ditches and tiling are methods of subsurface drainage that have been used successfully to control the level of the water table in many parts of the world.

Soil salt and sodium contents need to be measured to precisely determine the severity of the problem. The salt content of the soil is estimated from an electrical conductivity measurement using a soil water extract, soil water slurry or soil paste. The sodium content of the soil is often measured on a soil water extract and expressed as the ratio between the sodium and calcium plus magnesium and given the term sodium adsorption ratio (SAR).

Soils can be monitored by soil sampling the surface layer (top 6 inches) on a periodic basis (every three to five years). The SAR of the soil samples will indicate if there is a buildup of sodium. Generally, soils with an SAR of 13 from the saturated extract will exhibit significant physical problems due to dispersal of clay particles. Usually a soil with an SAR of 6 or lower from the saturated extract will not have physical problems associated with dispersed clay. However, if periodic sampling indicates that the SAR is increasing, say from 6 to 9, then it may be time to consider corrective action.

Topography of the Field

Topography or the "lay of the land" has a large impact on whether a field can be irrigated. Relief is a component of topography that refers to the difference in height between the hills and depressions in the field. The topographic relief will affect the type of irrigation system to be used, the water conveyance system (ditches or pipes), drainage requirements and water erosion control practices. The shape and arrangement of topographic landforms and the type of surface waterway network will also influence irrigation management.

Slope

Slope is important to soil formation and management because of its influence on runoff, soil drainage, erosion, use of machinery, and choice of crops. Slope is the incline or gradient of a surface and is commonly expressed in percent. The percent slope is determined by measuring the difference in vertical elevation in feet over 100 feet of horizontal distance. For example, a 5 percent slope rises or falls 5 feet per 100 feet of horizontal distance.

In addition to the percent of slope, the shape of the slope is another important characteristic. A convex slope curves outward like the outside surface of a ball, a concave slope curves inward like the inside surface of a saucer, and a plane slope is like a tilted flat surface.

Slopes are described as simple or complex. Simple slopes have a smooth appearance with surfaces extending in one or perhaps two directions. For example, slopes on alluvial fans and foot slopes of river valleys are regarded as simple. Complex areas have short slopes which extend in several directions and consist of convex and concave slopes much like the knoll and pothole topography found on glacial till plains.

Simple slopes of 1% or less are commonly used for gravity (surface) irrigation. Simple and complex slopes greater than 1% should only be irrigated with sprinkler or drip systems. Centre pivot sprinkler irrigation systems can operate on slopes up to 15%, but simple slopes greater than 9% are not generally recommended.

To accommodate an irrigation application method such as gravity or sprinkler systems, the slope in a field can be modified by land smoothing. However, land smoothing may cause yield reductions for one to three growing seasons. The places where topsoil was removed are most likely to have yield reductions. Special management of these areas through increased fertilizer and organic matter applications may be required for accelerated recovery.

Irrigation Water Quality

The quality of some water is not suitable for irrigating crops. Irrigation water must be compatible with both the crops and soils to which it will be applied. The Soil and Water Environmental Laboratory in the NDSU soil science department provides soil and water compatibility recommendations for irrigation. Generally a water analysis and a legal description of the land proposed for irrigation are required before a recommendation can be made.

The quality of water for irrigation purposes is determined by its salt content. An analysis of water for irrigation should include the *cations*: calcium, magnesium, and sodium, and the *anions*: bicarbonate, carbonate, sulfate, and chloride. Some crops are sensitive to boron, so it is often included in the analysis.

Irrigation Water Classification

The two most important factors to look for in an irrigation water quality analysis are the Total Dissolved Solids (TDS) and the Sodium Adsorption Ratio (SAR). The TDS of a water sample is a measure of the concentration of soluble salts in a water sample and is commonly referred to as the *salinity* of the water. TDS is expressed in terms of the electrical conductivity (EC) and its units are either:

millimhos per centimetre (mmhos/cm),

deci-Siemens per meter (dS/m) or

micromhos per centimetre (mmhos/cm)

where:

1000 mmhos/cm = 1 mmho/cm = 1 dS/m

The SAR of a water sample is the proportion of sodium relative to calcium and magnesium. Since it is a ratio, the SAR has no units.

Laboratories that perform irrigation water analysis may provide a suitability classification based on a system developed at the U.S. Salinity Laboratory in California. This classification system combines salinity and sodicity. For example, a water sample classified as C3-S2 would have a high salinity rating and a medium sodium rating. The scale for sodicity is not constant because it depends on the level of salinity. For example, an SAR of 8 is in the S1 category if the salinity is from 100 to 300 mmhos/cm; S2 if the salinity is from 300 to 3000 mmhos/cm, and S3 if the salinity is greater than 3000 mmhos/cm. Much of the water in North Dakota is classified in the C2 to C3 salinity range and the S1 to S2 sodium hazard range. In general, any water with an EC greater than 2000 mmhos/cm or an SAR value greater than 6 is not recommended for continuous irrigation in North Dakota. In cases where sporadic irrigation is practiced (i.e. a particular piece of land is only irrigated one year out of three or more), lower quality water may be used. However, the lower quality water should not have an EC that exceeds 3000 mmhos/cm or an SAR greater than 10.

Calcium added to irrigation water can lower the SAR and reduce the harmful effects of sodium. The effectiveness of added calcium depends on its solubility in the irrigation water. Calcium solubility is controlled by both the source of the calcium (e.g. calcium carbonate, gypsum, calcium chloride) and also the concentration of other ions in the irrigation water. Compared to calcium carbonate and gypsum, calcium chloride additions will result in higher concentrations of soluble calcium and be the most effective at lowering irrigation water SAR. However, calcium chloride is considerably more expensive than calcium carbonate and calcium sulfate (gypsum).

Carbonates

Carbonate and bicarbonate ions in the water combine with calcium and magnesium to form compounds which precipitate out of solution. Removing calcium and magnesium increases the sodium hazard to the soil from irrigation water. The increased sodium hazard is often expressed as "adjusted SAR." The increase of "adjusted SAR" over the SAR is a relative indication of the increase in sodium hazard due to the presence of these ions.

Precipitation of carbonate minerals has not been observed to plug sprinkler systems in North Dakota, but these minerals can cause plugging in drip irrigation systems. To control this problem, the pH of the irrigation water is generally lowered by adding a mild acid.

Salinity

C1 - Low salinity water — can be used for irrigation with most crops on most soils with little likelihood that soil salinity will develop. Some leaching is required, but this occurs under normal irrigation practices except in soils of slow and very slow permeability.

C2 - Medium salinity water — can be used if a moderate amount of leaching occurs. In most cases plants with moderate salt tolerance can be grown without special practices for salinity control.

C3 - High salinity water — cannot be used on soils with moderately slow to very slow permeability. Even with adequate permeability, special management for salinity control may be required and plants with good salt tolerance should be selected.

C4 - Very high salinity water — is not suitable for irrigation under ordinary conditions, but may be used occasionally under very special circumstances. The soils must have rapid permeability, drainage must be adequate, irrigation water must be applied in excess to provide considerable leaching, and very salt tolerant crops should be selected.

Sodium

S1 - Low sodium water — can be used for irrigation on almost all soils with little danger of development of harmful levels of exchangeable sodium.

S2 - Medium sodium water — will present an appreciable sodium hazard in fine textured soils, especially under low leaching conditions. This water may be used on coarse textured soils with moderately rapid to very rapid permeability.

S3 - High sodium water — will produce harmful levels of exchangeable sodium in most soils and requires special soil management, good drainage, high leaching, and high organic matter additions.

S4 - Very high sodium water — is generally unsatisfactory for irrigation purposes except at low and perhaps medium salinity.

Boron

Boron is essential for the normal growth of all plants, but the quantity required is very small. Plants sensitive to boron, such as dry beans, require much smaller amounts than plants that are tolerant of boron, such as corn, potatoes and alfalfa. In fact, the concentration of boron that will injure the sensitive plants is often close to that required for normal growth of tolerant plants.

Although there have been no documented problems with boron in water used for irrigation in North Dakota, testing for this element in irrigation water is a precautionary practice. Boron does occur in some North Dakota ground water at concentrations that are theoretically toxic to some crops. Boron concentration greater than 2 parts per million (ppm) may be a problem for certain sensitive crops, especially in years that require large quantities of irrigation water.

The Interaction Between Soil and Water

Soil is a medium that stores and moves water. If a cubic foot of a typical silt loam topsoil were separated into its component parts, about 45% of the volume would be mineral matter (soil particles), organic residue would occupy about 5% of the volume, and the rest would be *pore space.* The pore space is the voids between soil particles and is occupied by either air or water. The quantity and size of the pore spaces are determined by the soil's texture, bulk density and structure.

Water is held in soil in two ways: as a thin coating on the outside of soil particles and in the pore spaces. Soil water in the pore spaces can be divided into two different forms: gravitational water and capillary water. Gravitational water generally moves quickly downward in the soil due to the force of gravity. Capillary water is the most important for crop production because it is held by soil particles against the force of gravity.

As water infiltrates into a soil, the pore spaces fill with water. As the pores are filled, water moves through the soil by gravity and capillary forces. Water movement continues downward until a balance is reached between the capillary forces and the force of gravity. Water is pulled around soil particles and through small pore spaces in any direction by capillary forces. When capillary forces move water from a shallow water table upward, salts may precipitate and concentrate in the soil as water is removed by plants and evaporation.

Water Holding Capacity of Soils

There are four important levels of soil moisture content that reflect the availability of water in the soil. These levels are commonly referred to as: 1) saturation, 2) field capacity, 3) wilting point and 4) oven dry. When a soil is saturated, the soil pores are filled with water and nearly all of the air in the soil has been displaced by water. The water held in the soil between saturation and field capacity is gravitational water. Frequently, gravitational water will take a few days to drain through the soil profile and some can be absorbed by roots of plants.

Field capacity is defined as the level of soil moisture left in the soil after drainage of the gravitational water. Water held between field capacity and the wilting point is available for plant use.

The wilting point is defined as the soil moisture content where most plants cannot exert enough force to remove water from small pores in the soil. Most crops will be permanently damaged if the soil moisture content is allowed to reach the wilting point. In many cases, yield reductions may occur long before this point is reached.

Capillary water held in the soil beyond the wilting point can only be removed by evaporation. When soil is dried in an oven, nearly all water is removed. "Oven dry" moisture content is used to provide a reference for measuring the other three soil moisture contents.

When discussing the water holding capacity associated with a particular soil series, the water available for plant use in the *root zone* is commonly given (Table 3). Available soil water content is commonly expressed as inches per foot of soil. For example, the water available can be calculated for a soil with fine sandy loam in the first foot, loamy sand in the second foot and sand in the third foot. The top foot would have about 2.0 inches, the second foot would have about 1.0 inch and the third foot would have about 0.75 inches for a total of 3.75 inches of available water for a crop with a 3 foot root depth.

Table: *Available Soil Moisture Holding Capacity for Various Soil Textures.*

	Available Soil Moisture	
Soil Texture	***inches/inch***	***inches/foot***
Coarse Sand and Gravel	0.02 to 0.06	0.2 to 0.7
Sands	0.04 to 0.09	0.5 to 1.1
Loamy Sands	0.06 to 0.12	0.7 to 1.4
Sandy Loams	0.11 to 0.15	1.3 to 1.8
Fine Sandy Loams	0.14 to 0.18	1.7 to 2.2
Loams and Silt Loams	0.17 to 0.23	2.0 to 2.8
Clay Loams and Silty Clay Loams	0.14 to 0.21	1.7 to 2.5
Silty Clays and Clays	0.13 to 0.18	1.6 to 2.2

Soil Moisture Tension

The degree to which water clings to the soil is the most important soil water characteristic to a growing plant. This concept is often expressed as soil moisture tension. Soil moisture tension is negative pressure and commonly expressed in units of *bars*. During this discussion, when soil moisture tension becomes more negative it will be referred to as "increasing" in value.

Thus, as soil moisture tension increases (the soil water pressure becomes more negative), the amount of energy exerted by a plant to remove the water from the soil must also increase. One bar of soil moisture tension is nearly equivalent to -1 atmosphere of pressure (1 atmosphere of pressure is equal to 14.7 pounds per square inch at sea level).

A soil that is saturated has a soil moisture tension of about 0.001 bars, or less, which requires little energy for a plant to pull water away from the soil. At field capacity most soils have a soil moisture tension between 0.05 and 0.33 bars. Soils classified as sandy may have field capacity tensions around 0.10 bars, while clayey soil will have field capacity at a tension around 0.33 bars. At field capacity it is relatively easy for a plant to remove water from the soil.

The wilting point is reached when the maximum energy exerted by a plant is equal to the tension with which the soil holds the water. For most agronomic crops this is about 15 bars of soil moisture tension. To put this in perspective, the wilting point of some desert plants has been measured between 50 and 60 bars of soil moisture tension.

The presence of high amounts of soluble salts in the soil reduces the amount of water available to plants. As salts increase in soil water, the energy expended by a plant to extract water must also increase, even though the soil moisture tension remains the same. In essence, salts decrease the total available water in the soil profile.

How Plants Get Water From Soil

Water is essential for plant growth. Without enough water, normal plant functions are disturbed, and the plant gradually wilts, stops growing, and dies. Plants are most susceptible to damage from water deficiency during the vegetative and reproductive stages of growth. Also, many plants are most sensitive to salinity during the germination and seedling growth stages.

Most of the water that enters the plant roots does not stay in the plant. Less than 1% of the water withdrawn by the plant is actually used in photosynthesis (i.e. assimilated by the plant). The rest of the water moves to the leaf surfaces where it transpires (evaporates) to the atmosphere. The rate at which a plant takes up water is controlled by its physical characteristics, the atmosphere and soil environment.

As water moves from the soil, into the roots, through the stem, into the leaves and through the leaf stomata to the air, it moves from a low water tension to a high water tension. The water tension in the air is related to its relative humidity and is always greater than the water tension in the soil.

Plants can extract only the soil water that is in contact with their roots. For most agronomic crops, the root distribution in a deep uniform soil is concentrated near the soil surface. Over the course of a growing season, plants generally extract more water from the upper part of their root zone than from the lower part.

Plants such as grasses, with a high root density per unit of soil volume, may be able to absorb all available soil water. Other plants, such as vegetables, with a low root density, may not be able to obtain as much water from an equal volume of the same soil. Vegetables are generally more sensitive to water stress than high root density agronomic crops such as alfalfa, corn, wheat and sunflower.

Crop Water Use

Crop water use, also called evapotranspiration or ET, is an estimate of the amount of water transpired by the plants and the amount of evaporation from the soil surface around the plants. A plant's water use changes with a predictable pattern from germination to maturity. All agronomic crops have a similar water use pattern.

However, crop water use can change from growing season to growing season due to changes in climatic variables (air temperature, amount of sunlight, humidity, wind) and soil differences between fields (root depth, soil water holding capacities, texture, structure, etc.).

Many years of research have produced a number of equations that allow accurate estimates of crop water use values to be calculated from measured daily weather variables. Accurate estimates of crop water use values can be calculated for all the major irrigated crops in North Dakota.

Knowledge of water use patterns during the different growth stages has a major influence on how an irrigation system is designed and managed. Failure to recognize the water use patterns of a crop may result in poorly managed water applications.

Crop water stress, fertilizer and pesticide leaching and increased pumping costs are just a few of the results of poor irrigation water management.

Soil Quality Considerations in the Selection of Sites for Aquaculture

Soil quality is an important factor in fish pond productivity as it controls pond bottom stability, pH and salinity of overlying water and concentrations of plant nutrients required for the growth of phytoplankton, which is the base of food chain of the fish.

A satisfactory site for constructing fish ponds is that where the soil is very deep, with low salinity levels and neutral pH, water infiltration is very low, mineralization of organic matter takes place rapidly, nutrients are adsorbed and released slowly over a long period. Moderately heavy textured soils because of their high surface area and surface charge density posses the above characteristics. Too heavier textured soils such as pure clay may not be satisfactory as they have very high adsorptive property and thereby act as a sink for nutrients like phosphorus which may not be easily released to the overlying water.

These soils may also give problems of developments of deep cracks when dry (on draining the ponds) thereby allowing seepage losses of water. In order to have a clear understanding of the various physicochemical and biological processes which are controlled by the soil in the ponds and to make decisions on the suitability of sites for aquaculture as well as effective managements of the soils for increased productivity of the ponds, one needs to have a good background knowledge of the nature and properties of the soil. This chapter presents basic information on soils, their characteristics and their applications to aquaculture.

The physicochemical characteristics of water as they affect aquaculture are discussed separately elsewhere. The science of aquaculture has many similarities with that of agriculture though it is of recent development compared to the latter.

Many of the ideas in aquaculture are derived from the experiences in agriculture. Agriculture deals with soil-plant-animal relationships whereas aquaculture deals with the interaction of relationships of soil-water-phytobiota/zoobiota-fish.

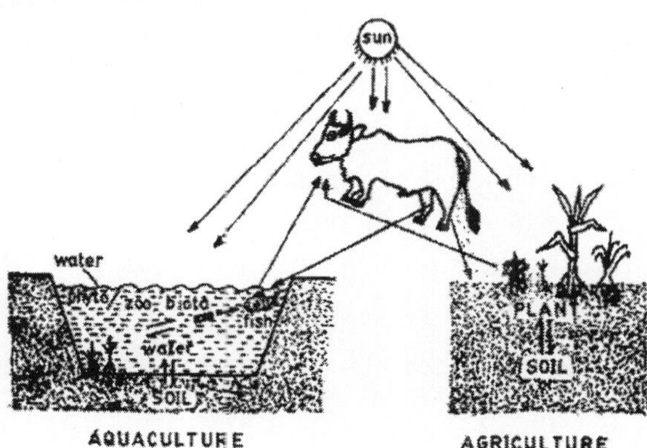

Figure: *Schematic comparaison of agriculture and aquaculture.*

Concepts of Soil

The concepts and viewpoints of soil differ according to its use. For example to a mining engineer the soil is the debris covering the rocks and minerals which he wishes to mine and therefore for him it is a nuisance and should be disposed of. To a highway engineer, soil is the materials on which roads are to be constructed. If the properties of the soil at a site are not suitable for road construction, he will replace them with gravels, rocks and other materials.

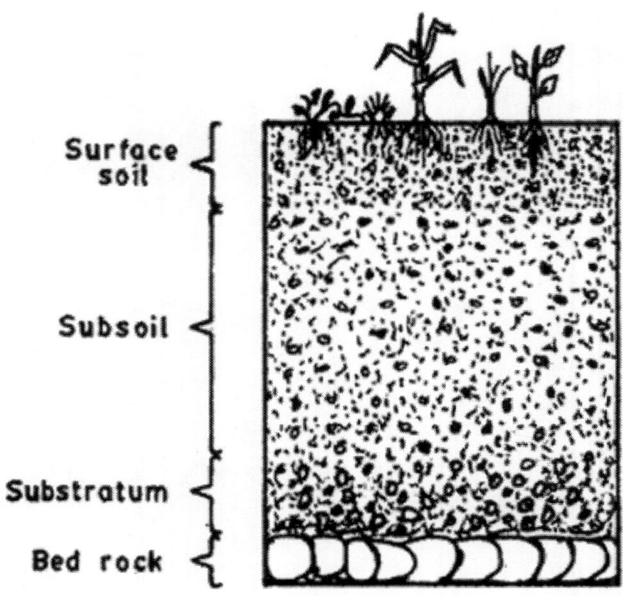

Figure: *Soil profile.*

An agriculturist or aquaculturist may view the soil as a natural body having depth, surface area and characteristic horizontal layers and formed at the earth's crest by the weathering of rocks and microbiological decomposition of organic residues.

In terms of soil's use, agriculturists consider soil as a medium where crops are cultivated whereas aquaculturists consider soil as a medium where ponds could be constructed and suitable soil-water conditions developed to culture fish and other aquatic organisms.

Composition of Soil

Soil is a complex body composed of five major components:

a. mineral matter obtained by the distintergration and decomposition of rocks;
b. organic matter, obtained by the decay of plant residues, animal remains and microbial tissues;
c. water, obtained from the atmosphere and the reactions in soil (chemical, physical and microbial);
d. air or gases, from atmosphere, reactions of roots, microbes and chemicals in the soil
e. organisms, both big (worms, insects) and small (microbes).

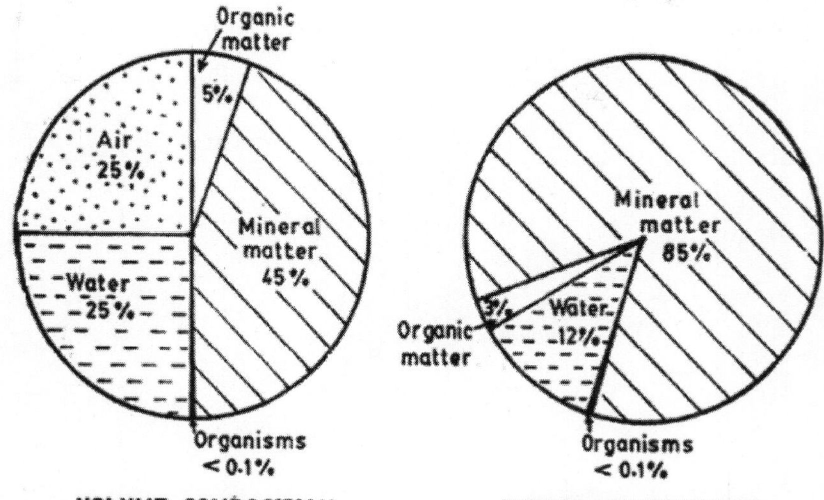

Figure: *Volume and weight composition of a soil (Percentage of air and water varies according to moisture saturation of soil).*

According to its size, soil can be separated into various fractions. Two common systems of classification are given in Table below.

Table: *Classification of soil particles according to two systems (U.S.D.A and International)*

Soil seperates	*U.S. Dept. of Agric. system Diameter (mm)*	*International system Diameter (mm)*	*Number of particles per g*
Very coarse sand	2.00 - 1.00	)) 2.00 - 0.20)	90
Coarse sand	1.00 - 0.50		720
Medium sand	0.50 - 0.25		5,700
Fine sand	0.25 - 0.10	)) 0.20 - 0.02)	46,000
Very fine sand	0.10 - 0.05		722,000
Silt	0.05 - 0.002	0.02 - 0.002	5,776,000
Clay	below 0.002	below 0.002	90,260,853,000

The clay fraction, because of its high surface area, is the most active part of the soil controlling many of the chemical and physical properties of the soil. It is the seat of soil fertility.

The sand and silt fractions i influence mainly the physical properties of the soil. The elemental composition of the inorganic component (mineral matter) of two soils formed from two types of rocks is presented in Table below. It could be seen from the table that Si, Al and Fe are the three major cations present in the soil.

Table: *Elemental composition (%) of two common rocks and soils developed from these rocks*

	Rocks ***Granite***	***Basalt***	***Soil*** ***FromGranite***	***From Basalt***
P_2O_5	0.13	0.34	0.08	0.13
K_2O	3.75	1.84	0.54	0.54
CaO	1.50	8.06	0.33	0.46
MgO	1.13	5.99	0.45	0.59
Sio_2	70.47	52.37	46.70	45.94
Al_2O_3	14.58	15.72	27.13	21.29
Fe_2O_3	3.57	9.88	12.20	19.31
Na_2O	2.99	3.28	0.40	0.13
TiO_2	0.61	1.24	1.73	1.92
MnO_2	0.07	0.15	0.04	0.37

The inorganic component (mineral matter) of the soil is composed of many types of minerals which influence the properties of the soil. The differences among soils are due mainly to the differences in the type and relative abundance of such minerals. Minerals are naturally occuring inorganic compounds having definite crystalline structures. They are classified into primary and secondary minerals.

Primary minerals are those formed at elevated temperature and inherited unchanged from igneous and metamorphic rocks whereas secondary minerals are formed at low temperature reactions and either inherited from sedimentary rocks or formed directly by weathering in soils. Some of the common types of minerals found in the various size fractions in the soil. It could be seen from the figure that the sand and silt fraction consists mainly of quartz and other primary minerals (feldspars, micas, pyroxene, olivine). In addition some small amounts of secondary minerals such as oxides of aluminium (gibbsite) and iron (hematite) are also found. The clay fraction is mainly composed of secondary silicate minerals such as kaolinite, illite and montmorillonite.

Soil organic matter could be considered to consist of two general groups: (i) fresh or partially decomposed plant and animal residues having some recognisable physical structures traceable to its origin; and (ii) the humus, which is a more resistant product of decomposition and colloidal in nature. The black or brown colour usually observed in the surface layers of soil profiles is due to the presence of humus. Humus is the most reactive part of the organic matter. Its capacity to hold water and nutrients greatly exceeds that of clay, its inorganic counterpart. The fresh and partially decomposed plant and animal residues generally occur in the sand and silt fraction of the soil and the humus occurs in the clay fraction.

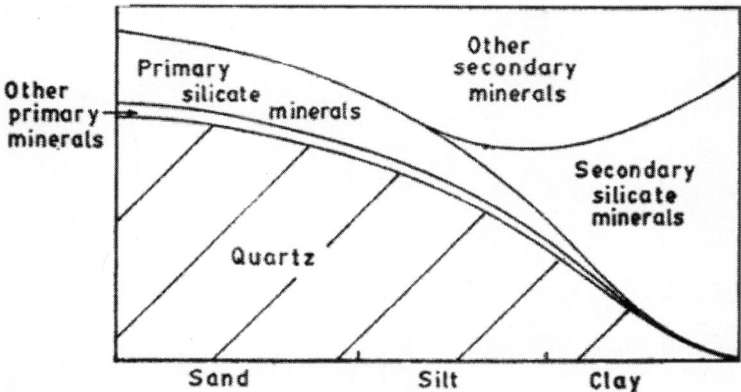

Figure: *Mineralogical composition of soil (area within the figure denotes the relative abundance of minerals).*

Physical Properties of the Soil

Texture

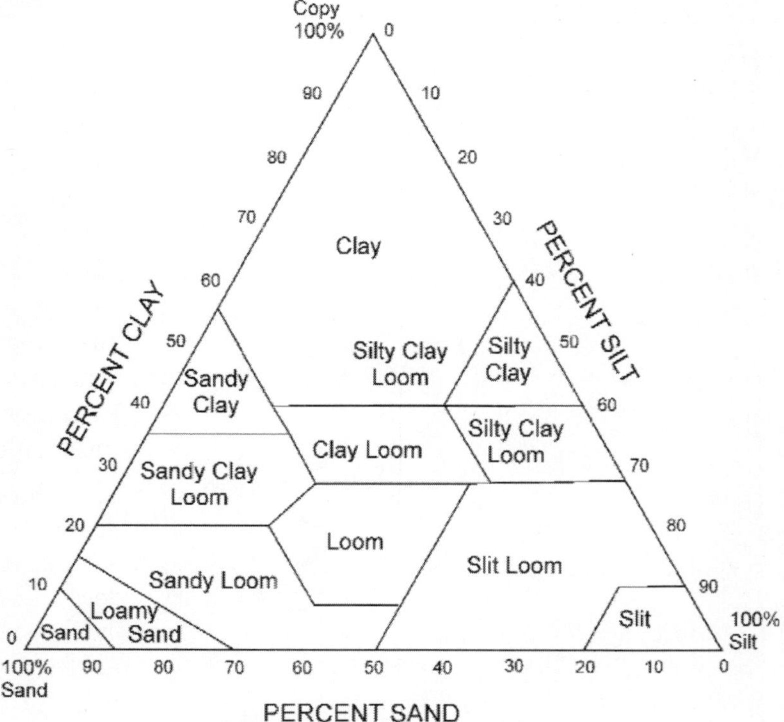

Figure: *Soil textural triangle.*

Texture refers to the relative proportions of particles of various sizes such as sand, silt and clay in the soil. The proportions of the separates in classes commonly used in describing soils are given in the textural triangle. In using the diagram, the points corresponding to the percentages of silt and clay present in the soil under consideration are located on the silt and clay lines respectively. Lines are then projected inward, parallel in the first case to the clay side of the triangle and in the second case parallel to the sand side. The name of the compartment in which the two lines intersect is the class name of the soil in question.

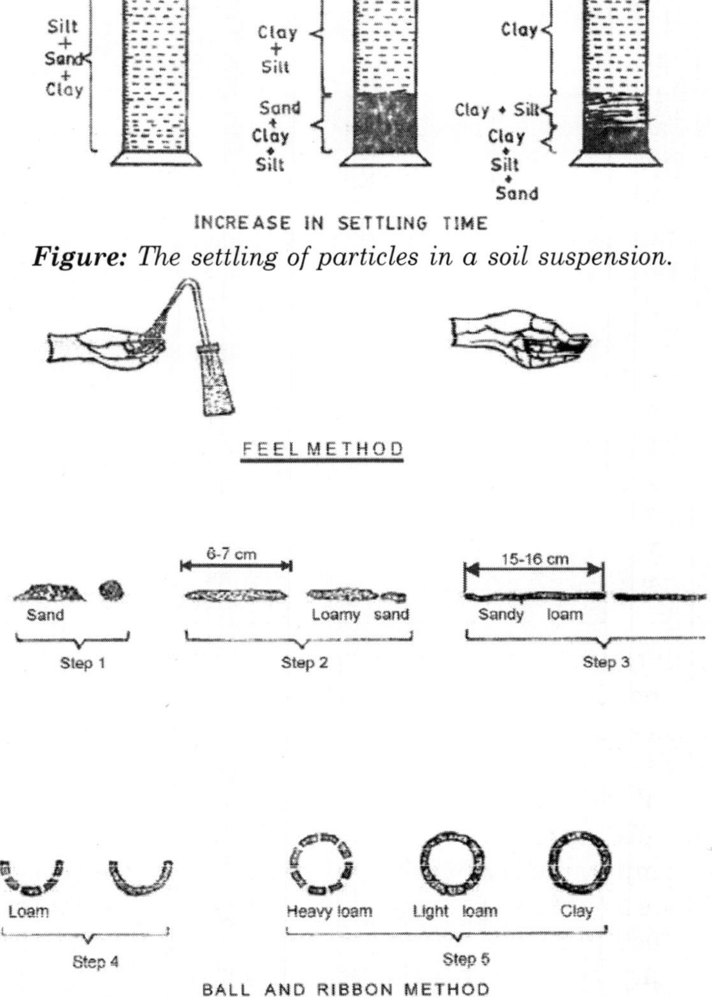

Figure: *The settling of particles in a soil suspension.*

Figure: *Field methods of soil texture estimation (Feel method and Ball and ribbon method).*

For examples a soil containing 15% clay, 20% silt and 65% sand is sandy loam and a soil containing equal amounts of sand, silt and clay is clay loam.

The percentages of sand, silt and clay in a soil could be determined in a soil laboratory by two standard methods - hydrometer method and pipette method. Both methods depend on the fact that at any given depth in a settling suspension the concentration of the particles varies with time, as the coarser fractions settle at a faster rate than the finer. In the field, soil texture could be estimated by the following methods.

(i) Feel method. In this method, the soil is moistened with water and rubbed between the thumb and fingers. The way the wet soil "slicks out" gives a good idea of the clay content. The sand particles are gritty, the silt has a floury or talcum - powder fell when dry and is only moderately plastic and sticky when wet. Accuracy of this method depends largely on experience.

(ii) Ball and ribbon method: The procedure of this method as described by Coche and Laughlin (1985) is as follows: Take a handful of soil and wet it so that it begins to stick together without sticking to the hand. A ball of about 3 cm diameter is made and put down. If it falls apart it is sand. If it sticks together roll the ball into a sausage shape 6 – 7 cm long. If it does not remain in this form it is loamy sand. If it remains in this shape, continue to roll until it reaches 15 – 16 cm long. If it does not remain in this form, it is sandy loam. If it remains in this shape, try to bend the sausage into a half circle and if it doesn't, it is a loam. If it does, bend the sausage to form a fullccircle and if it doesn't it is heavy loam. If it does with slight cracks in the sausage, it is light clay. If it does without any cracks, it is a clay.

(iii) Ball throwing method: The texture of the soil can be inferred by the way a ball of soil acts when it is thrown at a hard surface such as a wall or a tree. The steps to be followed in this method as described by Coche and Laughlin (1985) is as follows: Throw a ball of soil to a tree or wall 3 m away. If the soil is good only for splatter shots when either wet or dry, it has a coarse texture (loamy sand). If there is a "shot gun" pattern when dry and it holds its shape against medium range target when wet, it has a moderately coarse texture (sandy loam). If the ball shatters on impact when dry and clings together when moist but does not stick to the target it has a medium texture (loam, sandy clay loam, silty clay loam). If the ball holds its shape for long -

range shots when wet and sticks to the target but is fairly easy to remove it has a moderately fine texture (clay loam). If the ball sticks well to the target when wet and becomes a very hard missile when dry, it has a fine texture (clay).

The usual mechanical analysis of soils in the laboratory gives the percentages of the three size fractions, sand, silt and clay. For special uses, the same methods of laboratory analyses (pipette method or hydrometer method) can provide a much more detailed analysis giving further breakdown of the relative amounts of soil particles for more size classes in the form of a table or graph. The data in the graphical form is given as a particle - size frequency curve (PSF curve). PSF curves for selected soils.

The vertical axis represent the cumulative percentage of occurrance of the various particles sizes and the horizontal axis represents the logarithms of the particle size.

The vertical axis in the left hand side relate to the percentages of particles passing through sieves of a particular size and the vertical axis in the right hand side relate to the percentages of particles not passing through sieves of a particular size. The more vertical the PSF curve or part of the curve, the more uniform the particle size; a vertical line represents a perfect uniform particle size. The more inclined the curve or part of it, the greater the difference between the particle sizes (i.e. smaller porosity and higher compaction).

The inflexion point of the curve shows the most frequent particle size by weight. Fine textured soils have their curves towards the right hand side of the graph and the coarse textured soils to the left hand side. From the PSF curves, the percentages of silt, sand and clay can be calculated and using the textural triangular diagram the texture could be determined.

Soil texture is an important soil parameter determining the suitability of a site for aquaculture. A clayey soil stabilises pond bottom besides the fact that it adsorb large quantity of nutrients and release them slowly over a long period to the overlying water.

The clayey soil normally holds higher amounts of organic matter than light textured soils and thereby increase the productivity of the pond. It should be noted that too clay a soil (very sticky clay) may not be very satisfactory as it may give rise to fixation of phosphorus and create other physico-chemical biological problems. Such soils may give rise to cracks on draining the ponds, thereby increase seepage losses.

Structure

The term texture is used in reference to the size of individual soil particles but when the arrangement of the particles is considered the term structure is used. Structure refers to the aggregation of primary soil particles (sand, silt and clay) into compound particles or cluster of primary particles which are seperated by the adjoining aggregates by surfaces of weakness.

Structure modifies the effect of texture in regard to moisture and air relationships, availability of nutrients, action of microorganisms and root growth. E.g. a highly plastic clay (60% clay) is good for crop product if it has a well developed granular structure which facilitates aeration and water movement. Similarly a soil though has a heavy texture, can have a strongly developed structure, thus making it not very satisfactory for aquaculture as a result of this soil allowing high seepage losses.

Structure is defined in terms of grade, class and type of aggregates.

Grade: Grade of structure is the degree of aggregation and expresses the differential between cohesion within aggregates and adhesion between aggregates. These properties vary with the moisture content of the soil and it should be determined when the moisture content is normal - not when unusually dry or unusually wet. The four major grades of structure rated from 0 to 3 are listed below.

0. Structureless: no observable aggregation or no definite orderly arrangement of natural lines of weakness. Massive if coherent; single grain if noncoherent.
1. Weak: That degree of aggregation characterised by poorly formed indistinct aggregates that are barely observable in place. When disturbed, soil material that has this grade of structure breaks into a mixture of few entire aggregates many broken aggregates and much unaggregated material.
2. Moderate: Well formed distinct aggregates that are moderately durable and evident but not distinct in undisturbed soil. When disturbed, they break down into a mixture of many distinct entire aggregates, some broken aggregates and little unaggregated material.
3. Strong: Durable aggregates that are quite evident in undisturbed soil that adhere weakly to one another. When removed from the profile the sokl material consists very largely of entire aggreates and includes few broken ones and little or no nonaggregated material.

	Diameter of particle (mm)		
1	Gravel and sand(old alluvium)	A	Soil suitable for pond bottom if coefficient of permeability is less than 5×10^{-6} m/s
2	Sand		
3	Silt	B	Soil suitable for building dikes without impermeable clay core
4	Calcareous clayey soil (marl)	C	Soil suitable for pond bottom or dike only after modification of soil using amendments.
5	Heavy clay		

Class: The Class of structure describes the average size of individual aggregates and Type describe their form or shape. The various class divisions are: very fine or very thin, fine or thin, medium, coarse or thick and very coarse or very thick.

Water movement and drainage are poor in soils having blocky, prismatic, columnar and platy structures. These structured soils especially the platy type are most suitable for aquaculture.

Consistence is the resistance of a soil to deformation or rupture and is determined by the cohesive and adhesive properties of the soil mass. This is a term used to designate the manifestation of the cohesive and adhesive properties of soil at various moisture contents. A knowledge of the consistence of the soil is important in tillage operations, traffic and pond constructions. Consistence gives also an indication of the soil texture.

Consistence is described for three moisture levels:

1. Wet soil - non sticky, slightly sticky, sticky, very sticky; non plastic, slightly plastic, plastic and very plastic.
2. Moist soil - loose, very friable, friable, firm, very firm, extremely firm.
3. Dry soil - loose, soft, slightly hard, hard, very hard, extremely hard.

Description of the consistence terms mentioned above can be obtained from "Guidelines for Soil Profile Description" by FAO.

Partiole Density of soil is the mass per unit volume of soil particles (soil solid phase) - expressed in g/c.c. Most soils have particle density of about 2.6 g/cc. Presence of organic matter decrease the density and iron compounds increase the density.

Particle density can be determined using specific gravity bottle technique and bulk density by taking soil core samples of known volume in the field and determining the even dry weight. Water and air movements through soil depends on the pore space and the size distribution of the pores (microPores and macropores). Lower the pore space or higher the bulk density of the soil, the higher the suitability of the soil for aquaculture.

Atterberg Limits: From the previous section it could be noted that consistence of soils changes with the amount of moisture in the soil. Atterberg limits correspond to the moisture content at which a soil sample changes it's consistence from one state to the other. Liquid limit (LL) and plastic limit (PL) are two important states of consistence.

Liquid limit is the percentage moisture content at which a soil changes with decreasing wetness from the liquid to the plastic consistence or with increasing wetness from the plastic to the liquid consistence, whereas the plastic limit is the percentage moisture content at which a soil changes with decreasing wetness from the plastic to the semi-solid consistence or from the semi-solid to the plastic consistence.

Plastic index (PI) = LL - PL, is the moisture content range at which the soil remains plastic.

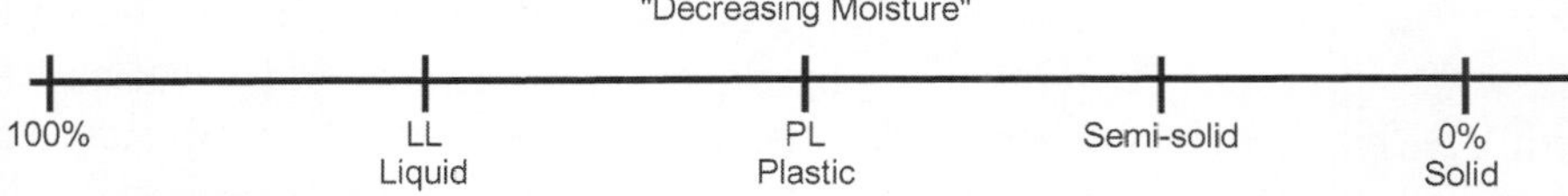

Table: *Typical laboratory tests showing average LL, PL and PI*

Soil type	***LL***	***PL***	***PI***
Sands	20	0	0
Silt	27	20	7
Clays	100	45	45
Colloidal clays	399	49	

According to Coche and Laughlin (1985) critical value of LL of the soil for constructing a pond dike without clay core, should be equal to at least 35% for best compaction results and for the constructing the impervious clay core of a pond dike soil material with LL smaller than 60%, PL smaller than 20% and PI more than 30% should be used.

Soil Colour: Soil colour gives an indication of the various processes going-on in the soil as well as the type of minerals in the soil. For example the red colour in the soil is due to the abundance of iron oxide under oxidised conditions (well-drainage) in the soil; dark colour is generally due to the accumulation of highly decayed organic matter; yellow colour is due to hydrated iron oxides and hydroxide; black nodules are due to manganese oxides; mottling and gleying are associated with poor drainage and/or high water table. Abundant pale yellow mottles coupled with very low pH are indicative of possible acid sulphate soils. Colours of soil matrix and mottles are indicative of the water and drainage conditions in the soil and hence suitability of the soil for aquaculture.

Soil colour is described by the parameters called hue, value and chroma. Hue represents the dominant wave length or colour of the light; value, refers to the lightness of the colour; chroma, relative purity or strength of the colour. The colour of the soil in terms of the above parameters could be quickly determined by comparison of the sample with a standard set of colour chips mounted in a note-book called MUNSELL SOIL COLOUR CHARTS. In these charts, the right hand top corner represents the Hue; the vertical axis, the value; and the horizontal axis, the chroma.

Soil Permeability: is the ability of the soil to transmit water and air. An impermeable soil is good for aquaculture as the water loss through seepage or infiltration is low. As the soil layers or horizons vary in their characteristics, the permeability also differs from one layer to another. Pore size, texture, structure and the presence of impervious layers such as clay pan determines the permeability of a soil. Clayey soils with platy structures have very low permeability.

Permeability is measured in terms of permeability rate or coefficient of permeability (cm per hour, cm per day, cm per sec.).

Permeability rate or coefficient of permeability is determined in the laboratory by measuring the rate of flow of water from a constant head of water through a colomn of soil at specific moisture content and other conditions. It is determined in the field by digging a hole of approximately 30 cm diameter, smearing the sides of the hole with heavy wet clay or lining with plastic sheet and measuring the rate of infiltration of water by filling the hole repeatedly with water and noting the time it takes for the water level to go down by a specific depth.

Coefficient of permeability of soils (pond bottom soils) suitable for aquaculture should be smaller than 5×10^{-6} m/s.

Chemical Properties Of Soil

Colloidal Properties

A major portion of the clay fraction (2mu) exists in the colloidal state (1 to 200 mu). This fraction exhibit the typical polloidal properties such as Tyndall effect, Brownian movement and possession of electric charges.

Ion Adsorption: The soil particles due to the presence of charges adsorb and exchange ions in solution. When fertilizers are added to agricultural soils or ponds, most of the nutrients in the fertilizer (cations as well as anions) are adsorbed on the negatively and positively charged sites of the soil or pond mud and released slowly to the soil water or pond water over a long period of time. This explains why fertilizers added to a pond may remain active for many years. In the case of phosphorus fertilizer, not much of the original fertilizer applied is washed out of the pond at each draining. The pond mud acts as a buffer system for many elements which could control the concentrations in the overlying waters because of the large concentrations of some elements present in the mud sediments. The effects of this buffer system could be to keep the concentrations in the overlying waters relatively constant even though the concentrations of the element in the water is altered. The large quantity of lime required to increase the pH of ponds is often due to the neutratization of the potential or exchange acidity resulting from adsorbed hydrogen and aluminium ions.

Origin of Charges in Soil and Ion Exchange

Electric charges on soil colloids arise from principally three sources:

i. from isomorphous substitution of one ion by another of different valency within the clay mineral structure. This gives rise to mainly negative charges. The charges are permanent and do not change with change in pH of the external solution.

ii. from ionisation of OH groups attached to the Al, Si, Fe at the edges of clay minerals. The charges created by this process are negative, zero or positive depending on the pH. Ionisation of OH groups attached to the Al at the edge of clay minerals can be represented as follows:

Clay - Al - OH^+_2	Clay - Al - OH	Clay - Al O^-
Very low pH	Intermediate pH	High pH
(below 3)	(about 3)	(above 3)

iii. The pH at which the soil has net zero charge, positive charge or negative charge depends on the type and amount of the various clay minerals present in the soil.

iv. from ionisation of - NH_2, - OH and - COOH functional groups in the organic matter of soils. This too gives rise to positive, neutral and negative charges as in (ii) depending on the pH.

Organic matter - $COOH^+_2$	Organic matter - COOH	Organic matter - COO^-
(pH < 2)	(pH < 2)	(pH < 2)

v. In a soil, charges arise from all the above 3 sources. Negative charges increase with increase in pH of the water surrounding the soil particles and positive charges increase with decrease in pH. At very low pH, the soil or pond mud adsorbs anions and acts as an anioh exchanger. At higher pH, it adsorbs cations and acts as a cation exchanger.

As the pH of most agricultural soils and ponds muds are generally higher than 4, the soil or mud is principally a cation exchanger, though it has some anion exchange properties. On the negatively charged sites or cation exchange sites, exchangeable cations are adsorbed and on the positively charged sites or anion exchange sites, exchangeable anions are adsorbed. Exchangeable cations can be acidic or basic cations. Acidic cations are those which produce acidity in soil. Common exchangeable cations and anions in soils are given below:

Exchangeable Cations

$$\left(\frac{Al^{3+}, H^{+}, Fe^{3+}}{\text{Acidic cations}}\right)\left(\frac{Ca^{2+}, Mg^{2+}, K^{+}, Na^{+}, Zn^{2+}}{\text{Basic cations}}\right) etc.$$

Exchangeable anions: $SO^{2-}{}_{4}$, $BO^{2-}{}_{3}$, $CO^{2-}{}_{3}$, HCO_3^-, OH^-, $H_2PO_4^-$, $HPO^{2-}{}_{4}$, $PO^{3}{}_{4}$, etc.

The quantity of cations which are adsorbed on the muds is expressed as milliequivalents of cations per 100 g (meq/100 g) of dry mud and is termed the cation exchange capacity (CEC). CEC is a measure of the total negative charges in the soil. CEC increase with increase in pH, % clay and % organic matter content in the soil. The fraction of CEC occupied by basic cations is called base saturation and the fraction of CEC occupied by acidic cations is base unsaturation. A sample calculation of CEC, base saturation and base unsaturation of a soil from Port Harcourt, Nigeria is given in Table.

Table: *CEC, base saturation and base unsaturation of a Typic Paleudult soil from Port Harcourt, Nigeria*

Exchangeable cations (meq/100g soil)						***CEC***	***Base***	***Base***
Ca^{2+}	Mg^{2+}	Na^{+}	K^{+}	Al^{3+}	H^{+}	meq/100g soil	saturation	unsaturation
1.9	0.6	0.8	0.2	1.2	0.8	5.5	0.64	0.36

Relationships between base unsaturation and soil pH and between base unsaturation and total hardness of water. Relative affinities of ions for adsorption on soils depend on the:

i. valency of the ions (higher the valency, higher the adsorptive power)
ii. concentration of the ions in the water (higher the concentration, higher the adsorptive power).

iii. nature of the ions (i.e. ions having same concentration and valency but different hydrated ionic radii have different affinities for adsorption)

e.g. $Li^+ < Na^+ < K^+ < Rb^+ < Cs^+$; $Ca^{2+} < Sr^{2+} < Ba^{2+}$

Lyotropic series (order of adsorptive power or replacing power). Ions having lower hydrated ionic radii have higher adsorptive power. Most ions are adsorbed at charged sites on the soil by electrostatic attraction (physical adsorption). Some ions are adsorbed by the formation of chemical bonds at neutral sites in the soil (chemical adsorption). The latter type of adsorption gives rise to fixation of the ion in the pond mud thereby the ion may not become available to the phytoplankton in the water for a long time. Here, the mud acts as a temporary sink for the nutrient.

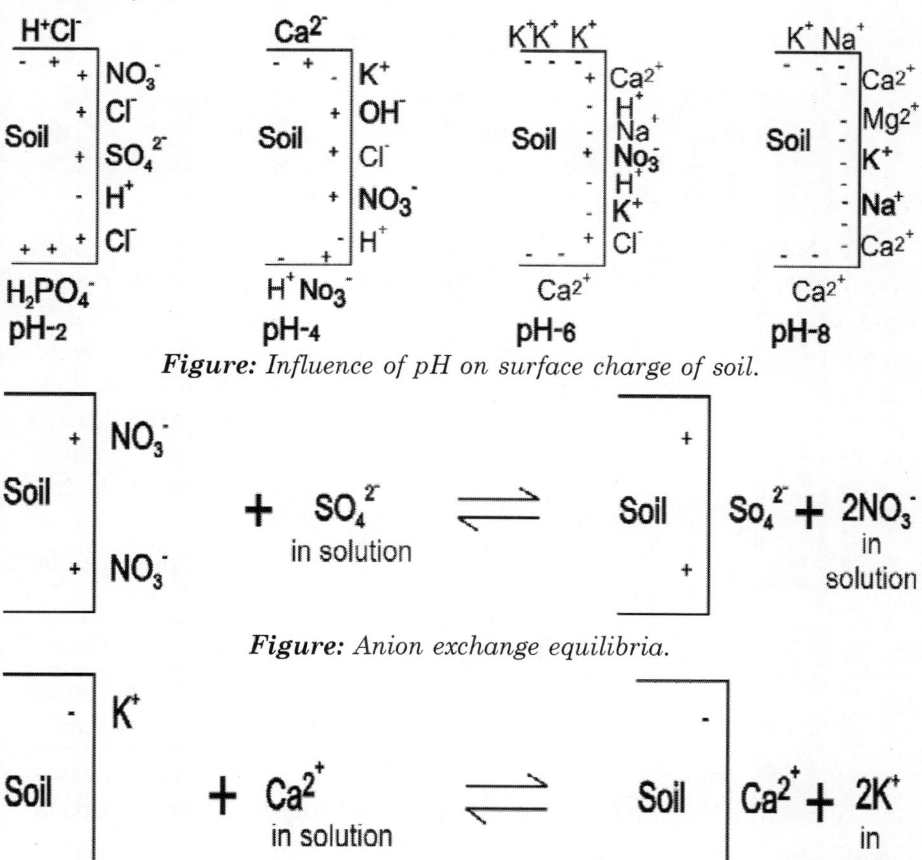

Figure: *Influence of pH on surface charge of soil.*

Figure: *Anion exchange equilibria.*

Figure: *Cation exchange equilibria.*

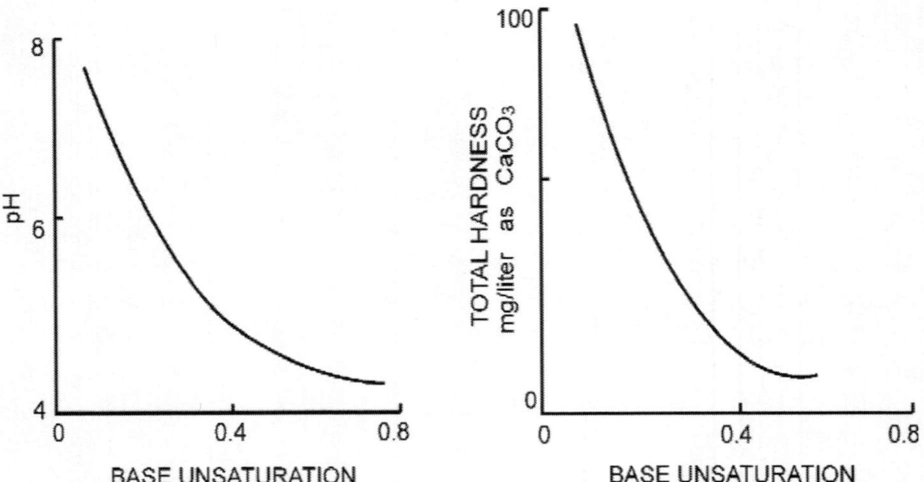

Figure 12: *Relationships of base unsaturation with pH and total hardness of water.*

When nutrients in the pond water are removed by phytoplankton or by any other process, the soil exchange complex releases more nutrients to the soil solution phase - the soil exchange complex acts as a store-house for these nutrients.

Similarly when lime or fertilizers are added to the pond water, a major portion of these materials get adsorbed on to the pond mud (Hickling, 1974).

Soil Acidity

A soil is said to be acidic if its pH is less than 7. Soil acidity can be further divided into:

	pH
Extremely acidic	<4.5
Very strongly acidic	4.5 – 5.0
Strongly acidic	5.1 – 5.5
Moderately acidic	5.6 – 6.0
Slightly acidic	6.1 – 6.5
Neutral	6.6 – 7.5

Acidity in the soil is caused by the following:

i. Adsorbed H^+, Fe^{3+} and Al^{3+} as well as solution H^+, Fe^{3+} and Al^{3+}. Hydrolyses of Fe^{3+} and Al^{3+} produces H^+ $Al^{3+} + H_2O = Al\ (OH)^{2+} + H^+$ $Al(OH)^{2+} + H_2O = Al\ (OH)^+_2 + H^+$ $Al(OH)^+_2 + H_2O = Al\ (OH)_3 + H^+$ precipitate

ii. Leaching of basic cations such as Ca, Mg, K and Na from the soil by heavy rainfall leaving acidic cations to remain in the soil. This is the reason for the occurance of acidic soils in humid regions and alkaline or neutral soils in arid regions.

iii. Oxidation of NH^+_4 and S^{2-} in soils by microorganisms.

$$NH^+_4 + 3O_2 = 2NO^-_2 + 4\ H^+ + H_2O$$

bacteria involved is nitrosomones

This reaction along with the following:

$$2NO^-_2 + O_2 = 2NO^-_3$$

bacteria involved is nitrobacter is called the nitrification process

Acid sulphate soils on aeration produce acidity by the oxidation of sulphide to sulphuric acid.

$$4\ Fe\ S_2 + 15O_2 + 2H_2O = 2\ Fe_2(SO_4)_3 + 2\ H_2SO_4$$

(pyrite)

bacteria involved is Thiobacillus ferroxidans

$$Fe(SO_4)_3 + Fe\ S_2 = 3\ Fe\ SO_4 + 2S$$

$$2S + 6\ Fe_2(SO_4)_3 + 8\ H_2O = 12\ Fe\ SO_4 + 8\ H_2SO_4$$

Oxidation of sulphur to sulphuric acid is caused by Thiobacillus thicxidans

$$S + 3/2\ O_2 + H_2O = H_2SO_4$$

The yellow mottles found in most acid sulphate soils are due to jarosite, a basic ferric sulphate compound $K\ Fe_3(SO_4)_2\ (OH)_6$ which is produced during pyrite oxidation with the hydrolysis of ferric sulphate and oxidation of ferrous sulphate.

$$Fe\ (SO_4)_3 + 2H_2O = 2\ Fe\ (OH)\ SO_4 + H_2SO_4$$

$$4\ Fe\ (SO_4) + 2\ H_2O + O_2 = 4\ Fe(OH)\ SO_4.$$

Soil acidity can be classified into two types:

i. active acidity

ii. potential acidity, exchange acidity or reserve acidity.

Active acidity is due to the Al^{3+} and H^+ (and to a less extent Fe^{3+}) in the water surrounding the soil particles and is measured by the pH meter. It is related to the degree of base unsaturation. Potential or exchange acidity is many times higher than the active acidity and is due to the adsorbed H^+ and Al^{3+}.

The magnitude of potential acidity is more dependent on CEC than the base unsaturation. Two soils with different CEC may have the same pH and base unsaturation but the soil with greater CEC will have the larger potential acidity.

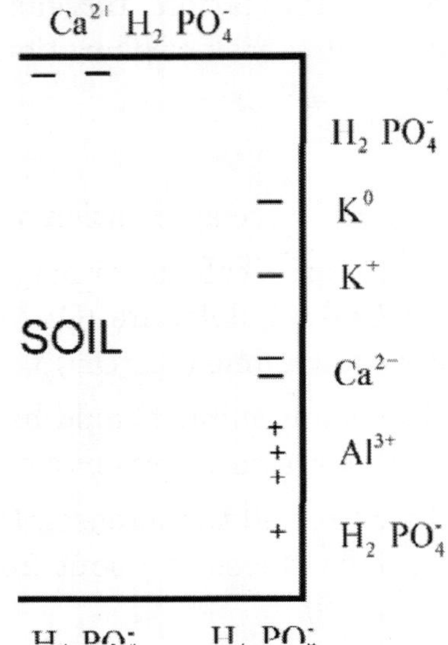

Figure: Phosphate fixation in soil (note adsorption of phosphate ions at neutral sites).

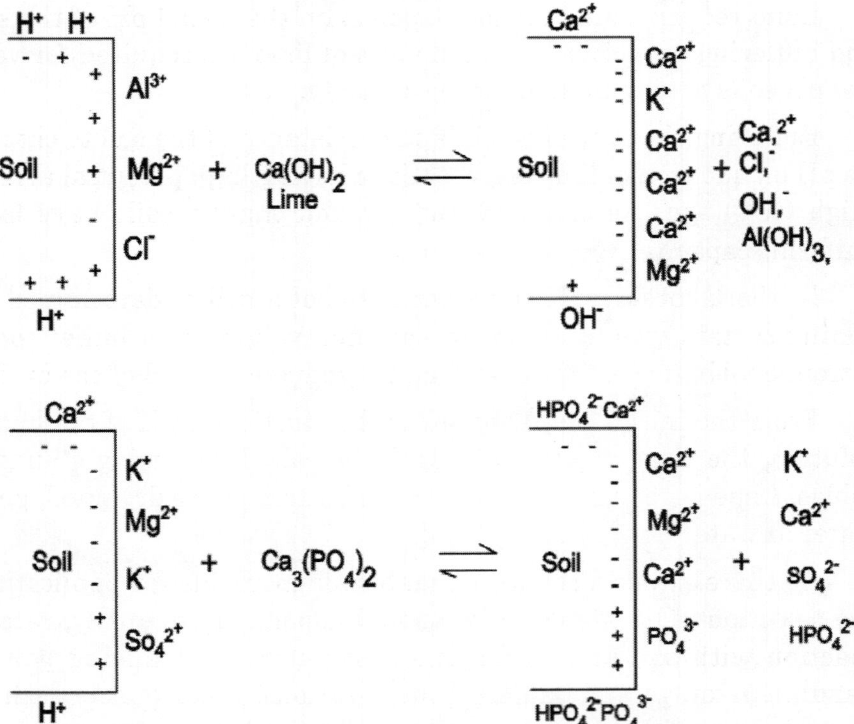

Figure: Reaction of lime and fertilizers with soil.

Acidity may give rise to Fe, Al, and Mn, toxicities, phosphorus unavailability (due to fixation), reduction in total hardness and total alkalinity, reduction in nitrification, increase in the amount of undecomposed organic matter, risk of liberation of certain parasitic and bacterial diseases.

Liming of the Soil

Soil acidity is corrected by the application of lime material.

The lime material has to be a calcium or magnesium salt of a weak acid such as limestone ($CaCO_3$), dolomite ($Ca\ Mg\ (CO_3)_2$), quicklime (CaO), hydrated lime or slaked lime ($Ca\ (OH)_2$).

In correcting acidity enough lime should be added to neutralize not only the active acidity but also the reserve or potential acidity.

Acidity is normally corrected to increase pH to about 5.9 (Boyd, 1979) At this pH, total hardness is about 20 mg/litre and base unsaturation 0.2 (Boyd, 1979).

Care should be taken not to overlime the soil as it would create deficiencies of phosphorus and many micronutrients.

Lime requirement of a soil depends on the initial pH of the soil and buffering capacity. Large amounts of lime are required for very low pH soils having high buffering capacity.

Buffering capacity of a soil is the resistance of the soil to change its pH on addition of lime or base. Soils having high potential acidity (high CEC) such as clay and high organic matter soils have high buffering capacity.

In the laboratory, lime requirement of a soil is determined by adding certain weight of soil to specified volume of a buffer (para nitrophenol buffer pH 8) and noting the reduction in pH of the buffer.

From the initial pH of the soil and reduction of pH of the buffer solution, the lime requirement could be calculated using standard tables. Lime is generally applied to the bottom of the dry pond, pond water or water flowing to the pond.

As there are several liming materials available for application, the selection of liming material should depend on its purity; rate of reaction with soil and water (fineness); cost; neutralising power; handling, storage and bagging considerations; other effects such as control of parasites, gill rot, precipitation of organic matter etc.

Table: *Neutralising power of lime materials*

Lime materials	*Molecular weight*	*Neutralising power (%)*
$Ca\ CO_3$	100	100
$Mg\ CO_3$	84	119
$Ca\ (OH)_2$	74	135
$Mg\ (OH)_2$	58	172
Ca O	56	178
Mg O	40	250

Note: Neutralising power is inversely proportional to the molecular weight.

Saline Soils

These are soils having excess soluble salts in the soil, that would impair with normal plant growth. If the salt content is too high, it may affect growth of phyto- and zoo-plankton. High concentrations of some specific elements like boron may also affect the growth of these organisms.

The salt content is normally measured by the specific conductivity of soil extract. If the specific conductivity of saturation soil paste extract is more than 4 m mhos/cm, most crops will not grow satisfactory. ppm salt is approximately equal to 640 × specific conductivity.

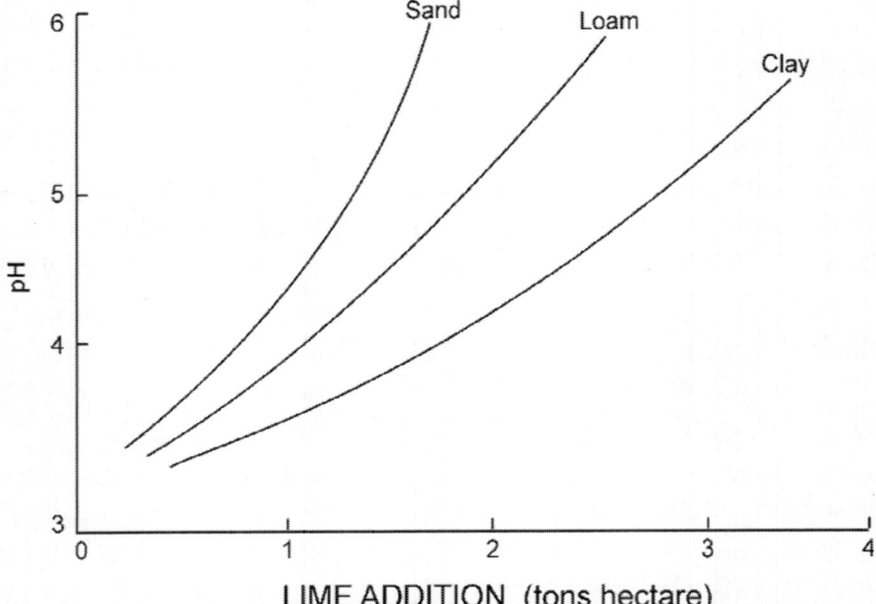

Figure: *Buffering capacities of soils with different textures.*

Saline soils generally occur in arid and semi-arid regions. It may also occur in localised areas in other regions due to poor drainage etc. These soils can be reclaimed by leaching the salts using good water and removing the leached salts by drains.

Sodic Soils

These are soils high in pH and exchangeable Na%- pH is generally higher than 8.5 and exchangeable Na occupies more than 15% of CEC. The poor drainage of these soils is not suitable for many upland crops but advantageous for aquaculture. The high Na content may affect the growth of phytoplankton, zooplankton and fish. These soils generally occur in arid and semi arid regions. They are reclaimed by treating the soil with gypsum ($Ca\ SO_4$) or sulphur.

Soil Fertility

Soil fertility is the ability of the soil to supply nutrients required by plants in adequate quantities and correct proportions. Plants require at least 16 elements to complete their life cycle. They are: C, H, O, N, P, K, Ca, Mg, S, Fe, Mn, Cu, Zn, Mo, B, Cl. Some of the lower plants in addition to the above elements require Co, V, Si. Among these C, H, O, N, P, K, Ca, Mg and S are required in large quantities and therefore called macronutrients and the others called micronutrients. The elements C, H, O are obtained mainly from air and water and the rest from the soil.

In a fertile soil, production is high at the start but diminish rapidly later due to exhaustion of the soil reserve of nutrients. In order to continue the high production of the pond, fertilizers containing the nutrients need to be applied frequently to the ponds. A good fraction of the fertilizers applied do not act directly on the organisms or water but often gets adsorbed on to the soil which releases the nutrients little at a time for a long period. As a result the fertilizers have prolonged action.

The amounts and kinds of fertilizers that need to be applied to the pond depend on the natural fertility of the soil. Fertility status of the soil can be determined by analysing samples of soils for nutrients in standard soil laboratories. For example the phosphorus fertility status of a soil can be determined by extracting phosphorus from the soil by a solution containing 0.03N NH_4F and 0.1 N HCl or by any other standard methods.

If the phosphorus extracted by the Bray and Kurtz No. 1 method is less than 10 ppm (10 ìg P per g soil) the soil is considered to be low in fertility and large quantity of phosphorus fertilizers need to be applied to the soil; if the value is 10 – 20 ppm, the soil is moderate with respect to phosphorus fertility and moderate quantity of phosphorus fertilizers need to be applied; if the soil has more than 20 ppm phosphorus, the soil is considered to be high in available phosphorus and no need to apply phosphorus fertilizers during the first year.

In a similar way, the fertility levels of other nutrients can be determined in the laboratory using appropriate methods for the elements concerned and rates of fertilizer requirements for the ponds can be determined.

Generalised ratings of the nutrient levels in agricultural soils are given in Table VI. The ratings may change slighly depending on the type of soil, ecology and the crop to be cultivated. However, they can be used as guides in interpreting soil test values. As there are no known ratings of soil nutrient values for aquaculture, the values given in Table VI. can be broadly used in determining the fertilizer needs of ponds. Types of fertilizers, their characterististics and methods of applications are discussed in a seperate chapter in this book. Nutrient cycles and transformations in the soil - water - organisms - atmosphere systems are also discussed elsewhere in this book.

Soil Formation, Soil Profile and Soil Classification

Soil Formation

Soils are formed by the weathering of rocks or materials deposited by rivers or wind. There are five groups of factors responsible for the kind, rate and extent of soil development. They are: Climate, organisms, parent material, topography and time. Soil from one place is different from another because of the differences in the influence of these factors.

The influence of climate is due to basically two factors: temperature and rainfall. Climate indirectly affects soil formation through its influence on organisms as well. High temperatures and rainfall increase the degree of weathering and therefore the extent of soil development.

Table: *Ratings of soil nutrient values in agricultural soils*

Nutrients	*very low*	*low*	*moderate*	*high*	*very high*
N (total N,%)	<0.05	0.05–0.15	0.15–0.20	0.20–0.30	>0.30
P (available P, Bray and Kurtz No.1, ppm)	<3	3 – 10	10 – 20	20 – 30	> 30
K (exchangeable K, meg/100g)	<0.2	0.2–0.3	0.3–0.6	0.6–1.0	> 1.0
Ca (exchangeable Ca meg/100g)	<2	2 – 5	5 – 10	10 – 20	> 20
Mg (exchangeable Mg, meg/100g)	<0.3	0.3–1	1 – 3	3 – 8	> 8

Increase of rainfall increase organic matter content, decrease pH, increase leaching of basic ions, movement of clay etc. Increase temperature increase organic matter decomposition and decrease its accumulation.

The influence of organisms is due to two factors: animal activities (humans, earthworm activity etc.) and influence of vegetation (grasslands produce soils very different from forests). Parent materials influence the type of minerals in the soil, nutrients, pH, texture etc. For example, soils formed on granite materials are more acidic and have less basic cations than those formed on basalt. Topography of the land influence the type of soil formation through its influence on drainage and slope.

Young soils do not have much horizon differentiation and would not have reached the advanced state of soil development compared to older soils. Among the 5 factors, climate has the greatest influence on soil formation. In areas of high rainfall and temperature, the soils formed are often similar even though the parent materials are different.

Soil Porfile

As the weathering of the parent materials progresses, with the various factors of soil formation acting on it, a more or less definite layering of the soil developes. A section from the surface through the various layers of the soil to the parent material is called the soil profile. The texture, depth, colour, chemical and physical properties, structure, and the sequence of the various horizons characterise a soil and determine its agricultural, aquacultural and other values.

The various layers are grouped into three headings A, B and C. The subdivisions are called horizons.

A group - (eluvial region) - zone of maximum leaching. Subdivided into Aoo, Ao, A_1 A_2, A_3

B group - (illuvial region) - zone of deposition of clay, Fe and Al oxides, Ca CO_3, Ca SO_4 (the last two materials generally occur in arid regions). B group is subdivided into B_1, B_2,

C group - unconsolidated mineral mass from which the A and B are developed.

Soil Classification

In order to have a systematic study of soils and to transfer knowledge from one area to the other the soils are classified according to their morphological, chemical and physical properties. As in the case of the classification of plants which are grouped into Class, Order,

Family, Genes, Species, the soils are also classified into categories such as Order, Suborder, Great Soil Group, Soil Series, Soil Types and Soil Phases. Order is divided into Zonal (determined primarily by climatic differences), Intrazonal (determined by difference in local conditions such as drainage, salinity which dominate the influence of climate) and Azonal (soils without horizon differentiation - alluvial deposite etc.).

Great Soil Group differs in the expressions of some specific condition in the soil profile. Soil Series have similar profile characteristics except for differences in the textures of the surface horizon. They are named normally by the name of the place where the series is first observed. Soil Types differ in the texture of the surface horizon. There are many types of classification systems, namely Soil Taxonomy (U.S.A.), French system (ORSTOM), Belgian system (INEAC), FAO - UNESCO Soil Legend system and many others. These classifications are developed primarily for agricultural applications. At the lower levels of classification (Order, Suborder, Great Soil Group) the various classification systems do not have much use in providing information on suitability of the soils for aquaculture. Classification at the Series or higher level is required in order to determine its suitability for aquaculture. Classification systems used for engineering purposes are generally based on particle size characteristics and same other soil physical properties like plasticity. One such classification system is the Unified Soil Classification (USC). There are three major soil units in the USC. They are:

i. Coarse - grained soils (CGS) which contain 50% or less of fines
ii. Fine - grained soils (FGS) which contain more than 50% fines
iii. Highly organic soils which are peat, muck, humus or swamp soil

Table: *Suitability of soils for construction of dikes and dams*

USC group symbol Coarse grained soils	***Descriptions***	***Suitability for dikes and dams***
GW	Well-graded gravels, gravel-sand mixtures, little or no fines	Very stable; permeable shells of dikes/ dam
GP	Poorly graded gravels, gravel and sand mixtures, little or no fines	Reasonably stable; permeable shells of dikes/dam
GM	Silty gravels, gravel and sand and silt mixture	Reasonably stable; not particularly suited to shells; but may be used for impermeable Cores or blankets
GC	Clayey gravel, gravel and sand and silt mixtures	Fairly stable; may be used for impermeable cores

Contd...

USC group symbol Coarse grained soils	***Descriptions***	***Suitability for dikes and dams***
SW	Well-graded sand, gravelly sands, little or no fines	Very stable; permeable sections; slope protection needed
SP	Poorly graded sands, gravelly sands, little or no fines	Reasonably stable; may be used
SM	Silty sands, sand and silt mixtures	Fairly stable; not particularly suited to shells; may be used for impermeable cores/dikes
SC	Clayed sands, sand and clay mixtures	Fairly stable; used for impermeable cores for flood control structures
Fine grained soils		
ML	Inorganic silts and very fine sands, rock flour, silty or clayey fine sands or clayey silts with slight plasticity	Poor stability; may be used for embankments with proper control
CL	Inorganic clays of low to medium plasticity, gravelly clays, sandy clays, silty clays and Lean clays	Stable; impermeable cores and blankets
OL	Organic silts and organic silty clays of low plasticity	Suitable for low embankments with very low hazard only
MH	Inorganic silts, micaceous or diatomaceous fine sandy or silty soils, elastic silts.	Poor stability; core of hydroulic fill dams; not desirable in rolled fill construction
CH	Inorganic clays of high plasticity, fat clays	Fairly stable with flat slopes; thin cores, blankets and dikes sections
OH	Organic clays of medium to high plasticity, organic silts	Suitable for low embankments with very low hazard only
Highly organic soils		
Pt	Peat and other organic soils	

The coarse and fine grained soils are subdivided according to liquid limit and plasticity index. Typical names and group symbols of USC system, their suitability for construction of dikes and dams are presented in Table VII. (Coche and Laughlin, 1985).

Soil Compaction

Soil compaction is understood as an increase of the state of compactness of a soil. The soil structure is made up by a skeleton of solid particles with different particle size varying from the coarse sandy to the fine clayey fraction. Between the solid particles are pore spaces that also vary in size and shape. The pore space is filled with air and water. It regulates the air and water regimes in soils. It conducts and stores water and air and it provides the channels for roots penetrating the soil. Soil compaction reduces the pore space and considerably impairs the properties of conductivity, storage and rootability. Compaction can be induced by natural processes as the transmission of clay perpendicular within the soil profile or by glacial pressure during the ice age or by human impacts as the pressure resulting from heavy machinery in agriculture and forestry.

The latter is actually of growing ecological concern, because the weights of machinery have been steadily grown in the past. The wheel loads of heavy machinery can reach 15 Mg in agriculture and 12 Mg in forestry. Soil compaction may have negative consequences for the productivity and the functionality of soils.

Experts estimate, that about 32% of the subsoils in Europe are highly susceptible and another 18% show moderate vulnerability to compaction. In agriculture especially subsoil compaction could become a severe long term threat to soil productivity and soil health.

For that reason human induced soil compaction should be watched and analysed in order to set up strategies to reduce or better to avoid it.

Ecological and Environmental Consequences of Soil Compaction

Soil compaction directly affects soil functions related to the productivity of soils. It has an impact on water and air storage capacity, oxygen supply to roots, and rootability. These properties influence plant growth and yield capacity. But also soil functions related to the environment, e.g. soil air and water conductivity, and heat balance are affected. Reduced water storage capacity and water conductivity can support flooding after heavy rainfall.

The reduced infiltration capacity may lead to a higher erosion susceptibility. Soil compaction induces changes in nutrient cycles, due to altered soil chemistry and greenhouse gas emissions can be elevated, due to alterations of the air regime of a soil. Since soil structure influences nearly all processes in the soil, compaction directly or indirectly affects the whole range of soil functions. The critical depth for harmful soil compaction in agricultural soils is the subsoil layer below the arable topsoil. In ploughed soils it is located in a depth of about 30 – 60 cm.

Chapter 7

Examination and Description of Soils

A description of the soils is essential in any soil survey. This chapter provides standards and guidelines for describing most soil properties and for describing the necessary related facts. For some soils, standard terms are not adequate and must be supplemented by a narrative. The length of time that cracks remain open, the patterns of soil temperature and moisture, and the variations in size, shape, and hardness of clods in the surface layer must be observed over time and summarized.

This chapter does not include a discussion of every possible soil property. For some soils, other properties need to be described. Good judgment will decide what properties merit attention in detail for any given pedon (sampling unit). Observations must not be limited by preconceived ideas about what is important.

Although the format of the description and the order in which individual properties are described are less important than the content of the description, a standard format has distinct advantages. The reader can find information more rapidly, and the writer is less likely to omit important features. Furthermore, a standard format makes it easier to code data for automatic processing. If forms are used, they must include space for all possible information.

Each investigation of the internal properties of a soil is made on a soil body of some dimensions. The body may be larger than a pedon or represent a portion of a pedon. During field operations, many soils are investigated by examining the soil material removed by a sampling tube or an auger. For rapid investigations of thin soils, a small pit can be dug and a section of soil removed with a spade. All of these are samples of pedons. Knowledge of the internal properties of a soil is derived mainly from studies of such samples. They can be studied more rapidly than entire pedons; consequently, a much larger number can be studied in many more places. For many soils, the information obtained from such a small sample describes the pedon from which it is taken with few omissions. For other soils, however, important properties of a pedon are not observable in the smaller sample, and detailed studies of entire pedons may be needed. Complete study of an entire pedon requires the exposure of a vertical section and the removal of horizontal sections layer by layer. Horizons are studied in both horizontal and vertical dimensions.

Some General Terms Used in Describing Soils

Several of the general terms for internal elements of the soil are described here; other more specific terms are described or defined in the following sections.

A *soil profile* is exposed by a vertical cut through the soil. It is commonly conceived as a plane at right angles to the surface. In practice, a description of a soil profile includes soil properties that can be determined only by inspecting volumes of soil. A description of a pedon is commonly based on examination of a profile, and the properties of the pedon are projected from the properties of the profile. The width of a profile ranges from a few decimeters to several meters or more. It should be sufficient to include the largest structural units.

A *soil horizon* is a *layer*, approximately parallel to the surface of the soil, distinguishable from adjacent layers by a distinctive set of properties produced by the soil-forming processes. The term layer, rather than horizon, is used if all of the properties are believed to be inherited from the parent material or no judgment is made as to whether the layer is genetic.

The *solum* of a soil consists of a set of horizons that are related through the same cycle of pedogenic processes. In terms of soil horizons described in this chapter, a solum consists of A, E, and B horizons and their transitional horizons and some O horizons. Included are horizons with an accumulation of carbonates or more soluble salts if they are either within, or contiguous, to other genetic horizons and are judged to be at least partly produced in the same period of soil formation. The solum of a soil presently at the surface, for example, includes all horizons now forming. It includes a bisequum (to be discussed). It does not include a buried soil or a layer unless it has acquired some of its properties by currently active soil-forming processes. The solum of a soil is not necessarily confined to the zone of major biological activity. Its genetic horizons may be expressed faintly to prominently. A solum does not have a maximum or a minimum thickness.

Solum and soils are not synonymous. Some *soils* include layers that are not affected by soil formation. These layers are not part of the solum. The number of genetic horizons ranges from one to many. An A horizon that is 10 cm thick overlying bedrock is by itself the solum. A soil that consists only of recently deposited alluvium or recently exposed soft sediment does not have a solum.

In terms of soil horizons described in this chapter, a solum consists of A, E, and B horizons and their transitional horizons and some O horizons. Included are horizons with an accumulation of carbonates or more soluble salts if they are either within, or contiguous, to other genetic horizons and are judged to be at least partly produced in the same period of soil formation.

The lower limit, in a general sense, in many soils should be related to the depth of rooting to be expected for perennial plants assuming that water state and chemistry are not limiting. In some soils the lower limit of the solum can be set only arbitrarily and needs to be defined in relation to the particular soil. For example, horizons of carbonate accumulation are easily visualized as part of the solum in many soils in arid and semiarid environments. To conceive of hardened carbonate accumulations extending for 5 meters or more below the B horizon as part of the solum is more difficult. Gleyed soil material begins in some soils a few centimetres below the surface and continues practically unchanged to a depth of many meters. Gleying immediately below the A horizon is likely to be related to the processes of soil formation in the modern soil. At great depth, gleying is likely to be relict or related to processes that are more geological than pedological. Much the same kind of problem exists in some deeply weathered soils in which the deepest material penetrated by roots is very similar to the weathered material at much greater depth.

For some soils, digging deep enough to reveal all of the relationships between soils and plants is not practical. Roots of plants, for example, may derive much of their moisture from fractured bedrock close to the surface. Descriptions should indicate the nature of the soil-rock contact and as much as can be determined about the upper part of the underlying rock.

A *sequum* is a B horizon together with any overlying eluvial horizons. A single sequum is considered to be the product of a specific combination of soil-forming processes.

Most soils have a single sequum, but some have two or more. A Spodosol, for example, can form in the upper part of an Alfisol, producing an eluviated zone and a spodic horizon underlain by another eluviated zone overlying an argillic horizon. Such a soil has two sequa. Soils in which two sequa have formed, one above the other in the same deposit, are said to be bisequal.

If two sequa formed in different deposits at different times, the soil is not bisequal. For example, a soil having an A-E-B horizon sequence may form in material that was deposited over another soil that already had an A-E-B horizon sequence. Each set of A-E-B horizons is a sequum but the combination is not a bisequum.

The lower set is a buried soil. If the horizons of the upper sequum extend into the underlying sequum, the affected layer is considered part of the upper sequum. For example, the A horizon of the lower soil may retain some of its original characteristics and also have some characteristics of the overlying soil. Here, too, the soils are not considered bisequal; the upper part of the lower soil is the parent material of the lower part of the currently forming soil. In many soils the distinction cannot be made with certainty. Nevertheless, the distinction is useful when it can be made. Where some of the C material of the upper sequum remains, the distinction is clear.

Studying Pedons

Pedons representative of an extensive mappable area are generally more useful than pedons that represent the border of an area or a small inclusion.

For a soil description to be of greatest value, the part of the landscape that the pedon represents and the vegetation should be described. This is referred to as the setting. The level of detail will depend on the objectives. A complete setting description should include information about the encompassing polypedon and, possibly, the polypedons conterminous with the encompassing polypedon. Furthermore, the setting may include information about the portion of the polypedon that differs from the central concept of the polypedon.

The description of a body of soil in the field, whether an entire pedon or a sample within it, should record the kinds of layers, their depth and thickness, and the properties of each layer. Generally, external features are observed throughout the extent of the polypedon; internal features are observed from the study of a pedon or that part of a pedon that is judged to be representative of the polypedon.

A pedon for detailed study of a soil is tentatively selected and then examined preliminarily to verify that it represents the desired segment of its range.

A pit exposing a vertical face approximately 1 meter across to an appropriate depth is satisfactory for most soils.

After the sides of the pit are cleaned of all loose material disturbed by digging, the exposed vertical faces are examined, usually starting at the top and working downward, to identify significant changes in properties. Boundaries between layers are marked on the face of the pit, and the layers are identified and described.

Photographs should be taken after the layers have been identified but before the vertical section is disturbed in the description-writing process. A point-count for estimation of the volume of stones or other features also is done before the layers are disturbed.

A horizontal view of each layer is useful. This exposes structural units that otherwise may not be observable. Patterns of colour within structural units, variations of particle size from the outside to the inside of structural units, and the pattern in which roots penetrate structural units are often seen more clearly in a horizontal section.

Excavations associated with roads, railways, gravel pits, and other soil disturbances provide easy access for studying soils; old exposures, however, must be used cautiously. The soils dry out or freeze and thaw from both the surface and the sides. Frequently, the soil structure in such excavations is more pronounced than is typical; salts may accumulate near the edges of exposures or be removed by seepage; and other changes may have taken place.

Depth to and Thickness of Horizons and Layers

Depth is measured from the soil surface. The soil surface is the top of the mineral soil; or, for soils with an 0 horizon, the soil surface is the top of the part of the 0 horizon that is at least slightly decomposed. Fresh leaf or needle fall that has not undergone observable decomposition is excluded from soil and may be described separately. The top of any surface horizon identified as an O horizon, whether Oi, Oe, or Oa, is considered the soil surface.

For soils with a cover of 80 percent or more rock fragments on the surface, the depth is measured from the surface of the rock fragments.

The depth to a horizon or layer boundary commonly differs within short distances, even within a pedon. The part of the pedon that is typical or most common is described. In the soil description, the horizon or layer designation is listed and is followed by the values that represent the depths from the soil surface to the upper and lower boundaries, in that order. The depth to the lower boundary of a horizon or layer is the depth to the upper boundary of the horizon or layer beneath it. The variation in the depths of the boundaries is recorded in the description of the horizon or layer. The depth limits of the deepest horizon or layer described include only that part actually seen.

In some soils the variations in depths to boundaries are so complex that usual terms for description of topography of the boundary are inadequate. These variations are described separately. For example, "depth to the lower boundary is mainly 30 to 40 cm, but tongues extend to depths of 60 to 80 cm." The lower boundary of horizon or layer and the upper boundary of the horizon or layer below share a common irregularity. The thickness of each horizon or layer is the vertical distance between the upper and lower boundaries. Thickness may vary within a pedon, and this variation should be shown in the description. A range in thickness may be given. It cannot be calculated from the range of upper and lower boundaries but rather must be evaluated across the exposure at different lateral points. The location of upper and lower boundaries are commonly in different places. The upper boundary of a horizon, for example, may range in depth from 25 to 45 cm and the lower boundary from 50 to 75 cm. Taking the extremes of these two ranges, a wrong conclusion could be that the horizon ranges in thickness from as little as 5 cm to as much as 50 cm.

Land Surface Configuration

Land surface configuration considered here is geometrical and includes *soil slope* and *land surface shape.* Landform from a morphogenetic aspect is not considered. It may be applicable to a pedon or to a larger area.

Land surface configuration and relief are quite different as used here, although the meanings may be similar in other contexts. *Relief,* in this context, refers to the elevation or differences in elevation above mean sea level, considered collectively, of a land surface on a broad scale. Elevation can be determined from topographic maps or by using a calibrated altimeter.

Soil Slope

Slope has a scale connotation. It refers to the ground surface configuration for scales that exceed about 10 meters and range upward to the landscape as a whole. Slope has gradient, complexity, length, and aspect. The scale of reference commonly exceeds that of the pedon and should be indicated. The scale may embrace a map unit delineation, component of it, or an arbitrary area.

Slope gradient is the inclination of the surface of the soil from the horizontal. It is generally measured with a hand level. The difference in elevation between two points is expressed as a percentage of the distance between those points. If the difference in elevation is 1 meter over a horizontal distance of 100 meters, slope gradient is 1 percent. A slope of 45° is a slope of 100 percent, because the difference in elevation between two points 100 meters apart horizontally is 100 meters on a 45° slope.

Overland flow gradient is the slope of the soil surface in the direction of flow of surface water if it were present. The following examples show equivalences between percentage gradient and degree of slope angle:

Percentage	*Angle*	*Angle*	*Percentage*
0	0°00'	0°	0
5	2°52'	2°	3.5
10	5°43'	4°	7.0
15	8°32'	6°	10.5
20	11°19'	8°	14.0
25	14°02'	10°	17.6
30	16°42'	12°	21.2
35	19°17'	15°	26.8
40	21°48'	20°	36.4
50	26°34'	25°	46.6
60	30°58'	30°	57.7
70	34°59'	35°	70.0
80	38°39'	40°	83.9
90	41°59'	45°	100.0
100	45°00'	50°	119.2

Slope Complexity refers to surface form on the scale of a mapping unit delineation. In many places internal soil properties are more closely related to the slope complexity than to the gradient. Slope complexity has an important influence on the amount and rate of runoff and on sedimentation associated with runoff.

A guide to terminology for various slope classes defined in terms of gradient and complexity. The terms are used in discussing soil slope, and they can also be used in naming slope phases.

Table: *Definitions of slope classes*

Classes	*Complex Slopes*	*Slope gradient limits*	
Simple Slopes		*Lower Percent*	*Upper Percent*
Nearly level	Nearly level	0	3
Gently sloping	Undulating	1	8
Strongly sloping	Rolling	4	16
Moderately steep	Hilly	10	30
Steep	Steep	20	60
Very steep	Very steep	>45	

Terms are provided for both simple and for complex slopes in some classes. Complex slopes are groups of slopes that have definite breaks in several different directions and in most cases markedly different slope gradients within the areas delineated.

Significance of slope gradient is tied to other soil properties and to the purposes of soil surveys.

Conventions are, therefore, provided in table 1 to adjust the slope limits of the various classes. Gently sloping or undulating soils, for example, can be defined to range as broadly as 1 to 8 percent or as narrowly as 3 to 5 percent. Classes may exceed the broadest range indicated in table 1 by a percentage point or two where the range is narrow and by as much as 5 percent or more where the range is broad.

If the detail of mapping requires slope classes that are more detailed than those in table 1, some of the classes can be divided as follows:

Nearly level: Level, Nearly level

Gently sloping: Very gently sloping, Gently sloping

Strongly sloping: Sloping, Strongly sloping, Moderately sloping

Undulating: Gently undulating, Undulating

Rolling: Rolling, Strongly rolling

In a highly detailed survey, for example, slope classes of 0 to 1 percent and 1 to 3 percent would be named "level" and "nearly level."

Slope length has considerable control over runoff and potential accelerated water erosion. Terms such as "long" or "short" can be used to describe slope lengths that are typical of certain kinds of soils. These terms are usually relative within a physiographic region. A "long" slope in one place might be "short" in another.

If such terms are used, they are defined locally. For observations at a particular point, it may be useful to record the length of the slope that contributes water to the point in addition to the total length of the slope. The former is called *point runoff slope length*. The *sediment transport slope length* is the distance from the expected or observed initiation upslope of runoff to the highest local elevation where deposition of sediment would be expected to occur. This distance need not be the same as the point runoff slope length.

Slope aspect is the direction toward which the surface of the soil faces. The direction is expressed as an angle between 0 degree and 360 degrees (measured clockwise from true north) or as a compass point such as east or north-northwest. Slope aspect may affect soil temperature, evapotranspiration, and winds received.

Land Surface Shape

Land surface shape has two components. One component is in a direction roughly parallel to the contours of the landform (or the contour lines on a map) as seen from directly overhead. The other component of shape is a direction perpendicular to the contours; that is, the shape of the slope as seen from the side. The shape parallel to the contours is less commonly consistent for a soil than is the shape perpendicular to the contours.

The shape parallel to the contours (across the slope) can be described by the shape of the contours. The shape is linear if contours are substantially a straight line, as on the side of a lateral moraine. An alluvial fan has a convex contour, as does a spur of the upland projecting into a valley. A cove on a hillside or a cirque in glaciated landscapes has concave contours. The two upper blocks have concave contours and the two lower blocks have convex contours. Where the contour is convex, runoff water tends to spread laterally as it moves down the slope. Where the contour is concave, runoff water tends to be concentrated toward the middle of the landform.

The shape of the surface at right angles to the contours (up and down the slope) may also be described as linear, convex, or concave. Shape in this direction is usually identified simply as slope shape in contrast to slope contour in the other dimension. The surface of a linear slope is substantially a straight line when seen in profile at right angles to the contours. The gradient neither increases nor decreases significantly with distance. An example is the dip slope of a cuesta. On a concave slope, gradient decreases down the slope as on foot slopes. Runoff water tends to decelerate as it moves down the slope, and if it is loaded with sediment, the water tends to deposit the sediment on the lower parts of the slope. The soil on the lower part of the slope also tends to dispose of water less rapidly than the soil above it.

On a convex slope, such as the shoulder or a ridge, gradient increases down the slope and runoff tends to accelerate as it flows down the slope. Soil on the lower part of the slope tends to dispose of water by runoff more rapidly than the soil above it. The soil on the lower part of a convex slope is subject to greater erosion than that on the higher part.

The configuration of the surface of a soil may be described in terms of both the shape of the contour and the shape of the slope. For example, a surface can be described as having a convex contour and a convex slope (an alluvial fan) or a linear contour and concave slope (the base of a moraine).

Microrelief refers to differences in ground-surface height, measured over distances of meters. Naturally formed features contrast with those that are tillage-determined. In areas of similar relief, the surface may be nearly uniform, or it may be interrupted by mounds, swales, or pits. Examples include the microrelief created when trees are blown over, referred to as cradle-knoll microrelief. This consists of the knoll left by the earth that clung to the roots of the tree when it was uprooted and the depression from which it came. Coppice dunes form where windblown soil material accumulates around widely spaced plants in arid regions. Gilgai produced by expansion and contraction of soils is a form of microrelief.

Mima mounds and biscuit-scabland are other examples of microrelief, although individual mounds may cover 100 square meters or more. Descriptions should indicate whether mounds or depressions are closed, form a network, or are in a linear pattern. If mounds rest on a smooth surface, their size and spacing should be described. At a specific site within an area having microrelief, it is important to note whether a described pedon is at a high point, on a slope, in a depression, or at some combination of these places. Internal soil properties in mounds may be different from the properties in depressions.

Roughness refers to a ground surface configuration with a repeat distance between prominences of less than 50 cm and for areas less than about 10 m across. This scale applies to most tillage operations and affects aspects of land surface water flow such as detention, infiltration, runoff, and erosion. Roughness, as used here, pertains to the ground surface and includes rock fragments on the surface. It does not include vegetation. If vegetation is included, the fact should be indicated. Roughness along a line, referred to as *one-dimensional roughness*, can be measured more easily than can roughness for an area.

Area measurements, however, permit the separation of random and tillage-determined roughness. The orientation to which the observation of *one-dimensional roughness* pertains must be specified relative to the direction of surface runoff or of air movement. Position within the tillage-determined relief, if present, should be indicated for *one-dimensional roughness.* An example of such a position would be the non-traffic interrow in a tilled field. The *standard deviation* of the ground surface height is the primary descriptor. There are a number of approaches to the measurement of roughness, and those who are in agronomic disciplines should be consulted. The measurements depend on the variation in height from a levelled reference. Photographs may be used to illustrate the classes; placement in classes may be made directly from the photographs.

Vegetation

Correlations between vegetation and soils are made for three main purposes: (1) understanding soil genesis, (2) recognizing soil boundaries, and (3) making predictions from soil maps about the kind and amount of vegetation produced.

The principal kinds of plants present are listed in order of their abundance. In annual cropland, the plant or plants that have been grown should be recorded, including significant weeds. In forested areas, separate treatment is often necessary for forest trees, understory of small trees and shrubs, and the ground cover. Many soils in range have an overstory of shrubs or low trees. These are listed separately from the grasses, forbs, and other ground cover. An idea of the density of stand or plant cover, such as average canopy cover of trees or shrubs, should be given. The range in size of dominant species of trees can be given as "diameter breast height," if desired. Estimated percentage of the ground covered by grasses and forbs should be included.

Common names of the plants may be used, if such names are clear and specific. In areas where the plants are important for the use and interpretation of the soil map, the soil survey record should include both common and scientific names of plants.

If possible, the kinds and amounts of plants in the potential natural vegetation on a soil should be estimated. This vegetation is closely related to the soil and its genesis. Generally, a close relationship exists between native vegetation and kinds of soil, yet there are important exceptions. Observations of the growth of native vegetation and cultivated crops aid in recognizing soil boundaries and provide direct information about the behaviour of specific plants on different kinds of soil. Within fields of a single crop, differences of vigour, stand, or colour of the crop or of weeds commonly mark soil differences and are valuable clues to the location of soil boundaries.

By studying many sites of the same kind of soil under different land-use history, the potential plant community and principles of plant succession for that kind of soil can be ascertained, particularly if range and forestry specialists provide assistance. Farmers learn which crops do well and which do poorly on different kinds of soil and adjust their cropping patterns accordingly. If the differences are large—as between crop failure and reasonable performance—the near absence of a given crop on a specific kind of soil questions the suitability of that kind of soil for the crop. If the differences are small, many non-soil factors can determine the farmer's choice of fields for a given crop. Yield information for cultivated crops, range, and trees should be associated with pedon descriptions insofar as possible.

Ground Surface Cover

The ground surface of most soils is covered to some extent at least part of the year by vegetation. Furthermore, in many soils rock fragments form part of the mineral material at the soil surface. Together, the vegetal material that is not part of the surface horizon and the rock fragments form the ground surface cover. The proportion of cover, together with its characteristics, is very important in determining thermal properties and resistance to erosion.

At one extreme, estimation of cover can be made visually without quantitative measurement. At the other extreme, transect techniques can be used to make a rather complete modal analyses of the ground surface. More effort is justified on ground surface documentation if it is relatively permanent. In many instances, a combination of rapid visual estimates and transect techniques is appropriate.

The ground surface may be divided into fine earth and material other than fine earth. The latter consists of rock fragments and both alive and dead vegetation. Vegetation is separated into *canopy* and *noncanopy*. A canopy component has a relatively large cross sectional area capable of intercepting rainfall compared to the area near enough to the ground surface to affect overland water flow. In practice, the separation of canopy from noncanopy should be coordinated with the protocols for computation of susceptibility to erosion. Noncanopy material is commonly referred to as *mulch*. It includes rock fragments and vegetation. The first step in evaluation is to decide upon the ground surface cover components. The number is usually one to three. A common three-component land surface consists of trees, bushes, and areas between the two. The areal proportion of each component must be established. This may be done by transect. If a canopy component is present, the area within the drip line as a percent of the ground surface is determined.

For each canopy component, the effectiveness must be established. *Effectiveness* is the percent of vertical raindrops that would be intercepted. Usually the canopy effectiveness is estimated visually, but a spherical densitometer may be used. In addition to the canopy effectiveness, the mulch (rock fragments plus vegetation) must be established for each component.

Transect techniques may be employed to determine the mulch percentage. The mulch can be subdivided into rock fragments and vegetation. From the areal proportions of the components and their respective canopy efficiencies and mulch percentages, the soil-loss ratio may be computed for the whole land surface (Wischmeier, 1978). In addition to the observations for the computation of the soil-loss ratio, information may be obtained about the percent of kinds of plants, size of rock fragments, amount of green leaf area, and aspects of colour of the immediate surface that would affect absorption of radiant energy in an area.

Parent Material

Parent material refers to unconsolidated organic and mineral materials in which soils form. The parent material of a genetic horizon cannot be observed in its original state; it must be inferred from the properties that the horizon has inherited and from other evidence. In some soils, the parent material has changed little, and what it was like can be deduced with confidence. In others, such as some very old soils of the tropics, the specific kind of parent material or its mode or origin is speculative.

Much of the mineral matter in which soils form is derived in one way or another from hard rocks. Glaciers may grind the rock into fragments and earthy material and deposit the mixture of particles as glacial till. On the other hand, rock may be weathered with great chemical and physical changes but not moved from its place of origin; this altered material is called "residuum from rock." In some cases, little is gained from attempting to differentiate between geologic weathering and soil formation because both are weathering processes.

It may be possible to infer that a material was weathered before soil formation. The weathering process causes some process constituents to be lost, some to be transformed, and others to be concentrated.

Parent material may not necessarily be residuum from the bedrock that is directly below, and the material that developed into a modern soil may be unrelated to the underlying bedrock. Movement of soil material downslope is an important process and can be appreciable even on gentle slopes, especially on very old landscapes. Also, locally associated soils may form in sedimentary rock layers that are different.

Seldom is there certainty that a highly weathered material weathered in place. The term "residuum" is used when the properties of the soil indicate that it has been derived from rock like that which underlies it and when evidence is lacking that it has been modified by movement. A rock fragment distribution that decreases in amount with depth, especially over saprolite, indicates that soil material probably has been transported downslope. Stone lines, especially if the stones have a different lithology than the underlying bedrock, provide evidence that the soil did not form entirely in residuum. In some soils, transported material overlies residuum and illuvial organic matter and clay are superimposed across the discontinuity between the contrasting materials. A certain degree of landscape stability is inferred for residual soils. A lesser degree is inferred for soils that developed in transported material.

Both consolidated and unconsolidated material beneath the solum that influence the genesis and behaviour of the soil are described in standard terms. Besides the observations themselves, the scientist records his judgment about the origin of the parent material from which the solum developed. The observations must be separated clearly from inferences.

The lithologic composition, structure and consistence of the material directly beneath the solum are important. Evidence of stratification of the material—textural differences, stone lines, and the like—need to be noted. Commonly, the upper layers of outwash deposits settled out of more slowly moving water and are finer in texture than the lower layers. Windblown material and volcanic ash are laid down at different rates in blankets of varying thickness. Examples of such complications are nearly endless.

Where alluvium, loess, or ash are rapidly deposited on old soils, buried soils may be well preserved. Elsewhere the accumulation is so slow that the solum thickens only gradually. In such places, the material beneath the solum was once near the surface but may now be buried below the zone of active change. Where hard rocks or other strongly contrasting materials lie near enough to the surface to affect the behaviour of the soil, their depths need to be measured accurately. The depth of soil over such nonconforming materials is an important criterion for distinguishing different kinds of soil.

Geological materials need to be defined in accordance with the accepted standards and nomenclature of geology. The accepted, authoritative names of the geological formations are recorded in soil descriptions where these can be identified with reasonable accuracy. As soil research progresses, an increasing number of correlations are being found between particular geological formations and the mineral and nutrient content of parent materials and soils. For example, certain terrace materials and deposits of volcanic ash that are different in age or source, but otherwise indistinguishable, vary widely in the content of cobalt. Wide variations in the phosphorus content of two otherwise similar soils may reflect differences in the phosphorus content of two similar limestones that can be distinguished in the field only by specific fossils.

Igneous rocks formed by the solidification of molten materials that originated within the earth. Examples of igneous rocks that weather to important soil material are granite, syenite, basalt, andesite, diabase, and rhyolite.

Sedimentary rocks formed from sediments laid down in previous geological ages. The principal broad groups of sedimentary rocks are limestone, sandstone, shale, and conglomerate. There are many varieties of these broad classes of sedimentary rocks; for example, chalk and marl are soft varieties of limestone. Many types are intermediate between the broad groups, such as calcareous sandstone and arenaceous limestone. Also included are deposits of diatomaceous earth, which formed, from the siliceous remains of primitive plants called diatoms.

Metamorphic rocks resulted from profound alteration of igneous and sedimentary rocks by heat and pressure. General classes of metamorphic rocks important as parent material are gneiss, schist, slate, marble, quartzite, and phyllite. The principal broad subdivisions of parent material are discussed in the following paragraphs.

Material Produced by Weathering of Rock in Place

The nature of the original rock affects the kinds of material produced by weathering. The rock may have undergone various changes, including changes in volume and loss of minerals—plagioclase feldspar and other minerals. Rock may lose mineral material without any change in volume or in the original rock structure, and saprolite is formed. Essentially, saprolite is a parent material. The point where rock weathering ends and soil formation begins is not always clear. The processes may be consecutive and even overlapping. Quite different soils may form from similar or even identical rocks under different weathering conditions. Texture, colour, consistence, and other characteristics of the material should be included in the description of soils, as well as important features such as quartz dikes. Useful information about the mineralogical composition, consistence, and structure of the parent rock itself should be added to help in understanding the changes from parent rock to weathered material.

Transported Material

The most extensive group of parent materials is the group that has been moved from the place of origin and deposited elsewhere. The principal groups of transported materials are usually named according to the main agent responsible for their transport and deposition. In most places, sufficient evidence is available to make a clear determination; elsewhere, the precise origin is uncertain.

In soil morphology and classification, it is exceedingly important that the characteristics of the material itself be observed and described. It is not enough simply to identify the parent material. Any doubt of the correctness of the identification should be mentioned.

For example, it is often impossible to be sure whether certain silty deposits are alluvium, loess, or residuum. Certain mud flows are indistinguishable from glacial till. Some sandy glacial till is nearly identical to sandy outwash. Fortunately, hard-to-make distinctions are not always of significance for soil behaviour predictions.

Material Moved and Deposited by Water

Alluvium.—Alluvium consists of sediment deposited by running water. It may occur on terraces well above present streams or in the normally flooded bottom land of existing streams. Remnants of very old stream terraces may be found in dissected country far from any present stream. Along many old established streams lie a whole series of alluvial deposits in terraces—young deposits in the immediate flood plain, up step by step to the very old deposits on the highest terraces. In some places recent alluvium covers older terraces.

Lacustrine deposits.—These deposits consist of material that has settled out of bodies of still water. Deposits laid down in freshwater lakes associated directly with glaciers are commonly included as are other lake deposits, including some of Pleistocene age that are not associated with the continental glaciers. Some lake basins in the Western United States are commonly called playas; the soils in these basins may be more or less salty, depending on climate and drainage.

Marine sediments.—These sediments settled out of the sea and commonly were reworked by currents and tides. Later they were exposed either naturally or following the construction of dikes and drainage canals. They vary widely in composition. Some resemble lacustrine deposits.

Beach deposits.—Beach deposits mark the present or former shorelines of the sea or lakes. These deposits are low ridges of sorted material and are commonly sandy, gravelly, cobbly, or stony. Deposits on the beaches of former glacial lakes are usually included with glacial drift.

Material Moved and Deposited by Wind

Windblown material can be divided into groups based on particle size or on origin. Volcanic ash and cinders are examples of materials classed by both particle size and origin. Other windblown material that is mainly silty is called loess, and that which is primarily sand is called eolian sand. Eolian sand is commonly but not always in dunes. Nearly all textures intermediate between silty loess and sandy dune material can be found.

Volcanic ash, *pumice*, and *cinders* are sometimes regarded as unconsolidated igneous rock, but they have been moved from their place of origin. Most have been reworked by wind and, in places, by water. Ash is volcanic ejecta smaller than 2 mm. Ash smaller than 0.05 mm may be called "fine ash." Pumice and cinders are volcanic ejecta 2 mm or larger.

Loess deposits typically are very silty but may contain significant amounts of clay and very fine sand. Most loess deposits are pale brown to brown, although gray and red colours are common. The thick deposits are generally massive and have some gross vertical cracking. The walls of road cuts in thick loess stand nearly vertical for years. Other silty deposits that formed in other ways have some or all of these characteristics. Some windblown silt has been leached and strongly weathered so that it is acid and rich in clay. On the other hand, some young deposits of windblown material (loess) are mainly silt and very fine sand and are low in clay. *Sand dunes*, particularly in warm, humid regions, characteristically consist of fine or medium sand that is high in quartz and low in clay-forming materials. Sand dunes may contain large amounts of calcium carbonate or gypsum, especially in deserts and semideserts.

During periods of drought and in deserts, local wind movements may mix and pile up soil material of different textures or even material that is very rich in clay. Piles of such material have been called "soil dunes" or "clay dunes." Rather than identify local accumulations of mixed material moved by the wind as "loess" or "dunes," however, it is better to refer to them as "wind-deposited material."

Also important but not generally recognized as a distinctive deposit is *dust*, which is carried for long distances and deposited in small increments on a large part of the world. Dust can circle the earth in the upper atmosphere. Dust particles are mostly clay and very fine silt and may be deposited dry or be in precipitation. The accumulated deposits are large in some places. An immense amount of dust has been distributed widely throughout the ages. The most likely sources at present are the drier regions of the world. Large amounts of dust may have been distributed worldwide during and immediately following the glacial periods.

Dust is an important factor affecting soils in some places. It is the apparent source of the unexpected fertility of some old, highly leached soils in the path of wind that blows from extensive deserts some hundreds of kilometers distant. It explains unexpected micronutrient distribution in some places. Besides dust, fixed nitrogen, sulfur, calcium, magnesium, sodium, potassium, and other elements from the atmosphere are deposited on the soil in varying amounts in solution in precipitation.

Material Moved and Deposited by Glacial Processes

Several terms are used for material that has been moved and deposited by glacial processes. *Glacial drift* consists of all of the material picked up, mixed, disintegrated, transported, and deposited by glacial ice or by water from melting glaciers. In many places glacial drift is covered by a mantle of loess. Deep mantles of loess are usually easily recognized, but very thin mantles may be so altered by soil-building forces that they can scarcely be differentiated from the underlying modified drift.

Glacial till.—This is that part of the glacial drift deposited directly by the ice with little or no transportation by water. It is generally an unstratified, heterogeneous mixture of clay, silt, sand, gravel, and sometimes boulders. Some of the mixture settled out as the ice melted with very little washing by water, and some was overridden by the glacier and is compacted and unsorted. Till may be found in ground moraines, terminal moraines, medial moraines, and lateral moraines. In many places it is important to differentiate between the tills of the several glaciations. Commonly, the tills underlie one another and may be separated by other deposits or old, weathered surfaces. Many deposits of glacial till were later eroded by the wave action in glacial lakes. The upper part of such wave-cut till may have a high percentage of rock fragments.

Glacial till ranges widely in texture, chemical composition, and the degree of weathering that followed its deposition. Much till is calcareous, but an important part is noncalcareous because no carbonate rocks contributed to the material or because subsequent leaching and chemical weathering have removed the carbonates.

Glaciofluvial deposits.—These deposits are material produced by glaciers and carried, sorted, and deposited by water that originated mainly from melting glacial ice. *Glacial outwash* is a broad term for material swept out, sorted, and deposited beyond the glacial ice front by streams of melt water. Commonly, this outwash is in the form of plains, valley trains, or deltas in old glacial lakes. The valley trains of outwash may extend far beyond the farthest advance of the ice. Near moraines, poorly sorted glaciofluvial material may form kames, eskers, and crevasse fills.

Glacial beach deposits.—These consist of rock fragments and sand. They mark the beach lines of former glacial lakes. Depending on the character of the original drift, beach deposits may be sandy, gravelly, cobbly, or stony.

Glaciolacustrine deposits.—These deposits are derived from glaciers but were reworked and laid down in glacial lakes. They range from fine clay to sand. Many of them are stratified or varved. A *varve* consists of the deposition for a calendar year. The finer portion reflects slower deposition during the cold season and the coarser portion deposition during the warmer season when runoff is greater.

Good examples of all of the glacial materials and forms described in the preceding paragraphs can be found. In many places, however, it is not easy to distinguish definitely among the kinds of drift on the basis of mode of origin and landform. For example, pitted outwash plains can scarcely be distinguished from sandy till in terminal moraines. Distinguishing between wave-cut till and lacustrine material is often difficult. The names themselves connote only a little about the actual characteristics of the parent material.

Material Moved and Deposited by Gravity

Colluvium is poorly sorted debris that has accumulated at the base of slopes, in depressions, or along small streams through gravity, soil creep, and local wash. It consists largely of material that has rolled, slid or fallen down the slope under the influence of gravity. Accumulations of rock fragments are called talus. The rock fragments in colluvium are usually angular, in contrast to the rounded, water-worn cobbles and stones in alluvium and glacial outwash.

Organic Material

Organic material accumulates in wet places where it is deposited more rapidly than it decomposes. These deposits are called peat. This peat in turn may become parent material for soils. The principal general kinds of peat, according to origin are:

Sedimentary peat. the remains mostly of floating aquatic plants, such as algae, and the remains and fecal material of aquatic animals, including coprogenous earth.

Moss peat. the remains of mosses, including Sphagnum.

Herbaceous peat. the remains of sedges, reeds, cattails, and other herbaceous plants.

Woody peat. the remains of trees, shrubs, and other woody plants.

Many deposits of organic material are mixtures of peat. Some organic soils formed in alternating layers of different kinds of peat. In places peat is mixed with deposits of mineral alluvium and/or volcanic ash. Some organic soils contain layers that are largely or entirely mineral material.

In describing organic soils, the material is called *peat* (fibric) if virtually all of the organic remains are sufficiently fresh and intact to permit identification of plant forms. It is called *muck* (sapric) if virtually all of the material has undergone sufficient decomposition to limit recognition of the plant parts. It is called *mucky peat* (hemic) if a significant part of the material can be recognized and a significant part cannot. Descriptions of organic material should include the origin and the botanical composition of the material to the extent that these can be reasonably inferred.

Contrasting Materials

Changes with depth that are not primarily related to pedogenesis but rather to geological processes are *contrasting soil materials* if they are sufficient to affect use and management. The term *discontinuity* is applied to certain kinds of contrasting soil materials.

Unconsolidated contrasting soil material may differ in pore-size distribution, particle-size distribution, mineralogy, bulk density, or other properties. Some of the differences may not be readily observable in the field. Some deposits are clearly stratified, such as some lake sediments and glacial outwash, and the discontinuities may be sharply defined.

Contrasting materials can be confused with the effects of soil formation. Silt content may decrease regularly with depth in soils presumed to have formed in glacial till. The higher silt content in the upper part of these soils can be explained by factors other than soil formation. In some of these soils, small amounts of eolian material may have been deposited on the surface over the centuries and mixed by insects and rodents with the underlying glacial till. In others, the silt distribution reflects water sorting.

Inferences about contrasting properties inherited from differing layers of geologic material may be noted when the soil is described. Generally, each identifiable layer that differs clearly in properties from adjacent layers is recognized as a subhorizon. Whether it is recognized as a discontinuity or not depends on the degree of contrast with overlying and underlying layers and the thickness. For many soils the properties inherited from even sharply contrasting layers are not consistent from place to place and are described in general terms. The C layer of a soil in stratified lake sediments, for example, might be described as follows: "consists of layers of silt and clay, 1 to 20 cm thick; the aggregate thickness of layers of silt and that of the layers of clay are in a ratio of about 4 to 1; material is about 80 percent silt."

Erosion

Erosion is the detachment and movement of soil material. The process may be natural or accelerated by human activity. Depending on the local landscape and weather conditions, erosion may be very slow or very rapid.

Natural erosion has sculptured landforms on the uplands and built landforms on the lowlands. Its rate and distribution in time controls the age of land surfaces and many of the internal properties of soils on the surfaces. The formation of Channel Scablands in the state of Washington is an example of extremely rapid natural, or geologic, erosion. The broad, nearly level interstream divides on the Coastal Plain of the Southeastern United States are examples of areas with very slow or no natural erosion.

Landscapes and their soils are evaluated from the perspective of their natural erosional history. Buried soils, stone lines, deposits of wind-blown material, and other evidence that material has been moved and redeposited is helpful in understanding natural erosion history. Thick weathered zones that developed under earlier climatic conditions may have been exposed to become the material in which new soils formed. In landscapes of the most recently glaciated areas, the consequences of natural erosion, or lack of it, are less obvious than where the surface and the landscape are of an early Pleistocene or even Tertiary age.

Even on the landscapes of most recent glaciation, however, postglacial natural erosion may have redistributed soil materials on the local landscape. Natural erosion is an important process that affects soil formation and, like man-induced erosion, may remove all or part of soils formed in the natural landscape.

Accelerated erosion is largely the consequence of human activity. The primary causes are tillage, grazing, and cutting of timber.

The rate of erosion can be increased by activities other than those of humans. Fire that destroys vegetation and triggers erosion has the same effect. The spectacular episodes of erosion, such as the soil blowing on the Great Plains of the Central United States in the 1930s, have not all been due to human habitation.

Frequent dust storms were recorded on the Great Plains before the region became a grain-producing area. "Natural" erosion is not easily distinguished from "accelerated" erosion on every soil. A distinction can be made by studying and understanding the sequence of sediments and surfaces on the local landscape, as well as by studying soil properties.

Landslip Erosion

Landslip erosion refers to the mass movement of soil. Slides and flows are two kinds of landslip erosion. In the slide process, shear takes place along one or a limited number of surfaces. Slide movement may be categorized as *slightly* or *highly deformed*, depending on the extent of rearrangement from the original organization. In flow movement the soil mass acts as a viscous fluid. Failure is not restricted to a surface or a small set of surfaces.

Classes of landslip erosion are not provided. Location of the mass movement relevant to landscape features generally and the size of the mass movement in terms of area parallel to the land surface and the depth may be indicated. Information about the time since the mass movement took place may be very useful.

Water Erosion

Water erosion results from the removal of soil material by flowing water. A part of the process is the detachment of soil material by the impact of raindrops. The soil material is suspended in runoff water and carried away. Four kinds of accelerated water erosion are commonly recognized: sheet, rill, gully, and tunnel (piping).

Sheet erosion is the more or less uniform removal of soil from an area without the development of conspicuous water channels. The channels are tiny or tortuous, exceedingly numerous, and unstable; they enlarge and straighten as the volume of runoff increases. Sheet erosion is less apparent, particularly in its early stages, than other types of erosion. It can be serious on soils that have a slope gradient of only 1 or 2 percent; however, it is generally more serious as slope gradient increases.

Rill erosion is the removal of soil through the cutting of many small, but conspicuous, channels where runoff concentrates. Rill erosion is intermediate between sheet and gully erosion. The channels are shallow enough that they are easily obliterated by tillage; thus, after an eroded field has been cultivated, determining whether the soil losses resulted from sheet or rill erosion is generally impossible.

Gully erosion is the consequence of water that cuts down into the soil along the line of flow. Gullies form in exposed natural drainage-ways, in plow furrows, in animal trails, in vehicle ruts, between rows of crop plants, and below broken man-made terraces. In contrast to rills, they cannot be obliterated by ordinary tillage. Deep gullies cannot be crossed with common types of farm equipment.

Gullies and gully patterns vary widely. V-shaped gullies form in material that is equally or increasingly resistant to erosion with depth. U-shaped gullies form in material that is equally or decreasingly resistant to erosion with depth. As the substratum is washed away, the overlying material loses its support and falls into the gully to be washed away. Most-U-shaped gullies become modified toward a V shape once the channel stabilizes and the banks start to spall and slump. The maximum depth to which gullies are cut is governed by resistant layers in the soil, by bedrock, or by the local base level. Many gullies develop headward; that is, they extend up the slope as the gully deepens in the lower part.

Tunnel erosion may occur in soils with subsurface horizons or layers that are more subject to entrainment in moving free water than is the surface horizon or layer. The free water enters the soil through ponded infiltration into surface-connected macropores. Desiccation cracks and rodent burrows are examples of macropores that may initiate the process. The soil material entrained in the moving water moves downward within the soil and may move out of the soil completely if there is an outlet. The result is the formation of tunnels (also referred to as pipes) which enlarge and coalesce. The portion of the tunnel near the inlet may enlarge disproportionately to form a funnel-shaped feature often referred to as a "jug." Hence, the term "piping" and "jugging." The phenomenon is favoured by the presence of appreciable exchangeable sodium.

Deposition of sediment carried by water is likely anywhere that the velocity of running water is reduced—at the mouth of gullies, at the base of slopes, along stream banks, on alluvial plains, in reservoirs, and at the mouth of streams. Rapidly moving water, when slowed, drops stones, then cobbles, pebbles, sand, and finally silt and clay. Sediment transport slope length has been defined as the distance from the highest point on the slope where runoff may start to where the sediment in the runoff would be deposited.

Wind Erosion

Wind Erosion in regions of low rainfall, can be widespread, especially during periods of drought. Unlike water erosion, wind erosion is generally not related to slope gradient. The hazard of wind erosion is increased by removing or reducing the vegetation.

When winds are strong, coarser particles are rolled or swept along on or near the soil surface, kicking finer particles into the air. The particles are deposited in places sheltered from the wind. When wind erosion is severe, the sand particles may drift back and forth locally with changes in wind direction while the silt and clay are carried away. Small areas from which the surface layer has blown away may be associated with areas of deposition in such an intricate pattern that the two cannot be identified separately on soil maps.

Estimating the Degree of Erosion

The degree to which accelerated erosion has modified the soil may be estimated during soil examinations. The conditions of eroded soil are based on a comparison of the suitability for use and the management needs of the eroded soil with those of the uneroded soil. The eroded soil is identified and classified on the basis of the properties of the soil that remains. An estimate of the soil lost is described. Eroded soils are defined so that the boundaries on the soil maps separate soil areas of unlike use suitabilities and unlike management needs.

The depth to a reference horizon or soil characteristic of the soil under a use that has minimized accelerated erosion are compared to the same properties under uses that have favoured accelerated erosion. For example, a soil that supports native grass or large trees with no evidence of cultivation would be the basis for comparison of the same or similar soil that has been cleared and cultivated for a relatively long time.

The depth to reference layers is measured from the top of the mineral soil because organic horizons at the surface of mineral soils are destroyed by cultivation.

The depths to a reference layer must be interpreted in terms of recent soil use or history. Cultivation may cause differences in thickness of layers. The upper parts of many forested soils have roots that make up as much as one-half of the soil volume. When these roots decay, the soil settles. Rock fragment removal can also lower the surface. The thickness of surficial zones that have been bulked by tillage should be adjusted downward to what they would be if water had compacted them.

The thickness of a plowed layer of a specific soil cannot be used as a standard for either losses or additions of material because, as a soil erodes, the plow cuts progressively deeper. Nor can the thickness of the uncultivated and uneroded A horizon be used as a standard for all cultivated soil, unless the A horizon is much thicker than the plow layer. If the horizon immediately below the plowed layer of an uneroded soil is distinctly higher in clay than the A horizon, the plow layer becomes progressively more clayey under continued cultivation as erosion progresses; the texture of the plow layer may then be a criterion of erosion.

Comparisons must be made on comparable slopes. Near the upper limit of the range of slope gradient for a soil, horizons may normally be thinner than near the lower limit of the range for the same soil.

Roadsides, cemeteries, fence rows, and similar uncultivated areas that are a small part of the landscape as a whole or are subject to unusual cultural histories must be used cautiously for setting standards, because the reference standards for surface-layer thickness are generally set too high. In naturally treeless areas or in areas cleared of trees, dust may collect in fence rows, along roadsides, and in other small uncultivated areas that are covered with grass or other stabilizing plants. The dust thus accumulated may cause the surface horizon to become several centimetres thicker in a short time.

For soils having clearly defined horizons, differences due to erosion can be accurately determined by comparison of the undisturbed or uncultivated norms within the limitations discussed.

Guides for soils having a thin A horizon and little or no other horizon are more difficult to establish. After the thin surface layer is gone or has been mixed with underlying material, few clues remain for estimating the degree of erosion. The physical conditions of the material in the plowed layer, the appearance and amount of rock fragments on the surface, the number and shape of gullies, and similar evidence are relied on. For many soils having almost no horizon expression, attempting to estimate the degree of erosion serves little useful purpose.

Classes of Accelerated Erosion

The classes of accelerated erosion that follow apply to both water and wind erosion. They are not applicable to landslip or tunnel erosion. The classes pertain to the proportion of upper horizons that have been removed. These horizons may range widely in thickness; therefore, the absolute amount of erosion is not specified.

Class 1. This class consists of soils that have lost some, but on the average less than 25 percent, of the original A and/or E horizons or of the uppermost 20 cm if the original A and/or E horizons were less than 20 cm thick. Throughout most of the area, the thickness of the surface layer is within the normal range of variability of the uneroded soil. Scattered small areas amounting to less than 20 percent of the area may be modified appreciably.

Evidence for class 1 erosion includes (1) a few rills, (2) an accumulation of sediment at the base of slopes or in depressions, (3) scattered small areas where the plow layer contains material from below, and (4) evidence of the formation of widely spaced, deep rills or shallow gullies without consistently measurable reduction in thickness or other change in properties between the rills or gullies.

Class 2. This class consists of soils that have lost, on the average, 25 to 75 percent of the original A and/or E horizons or of the uppermost 20 cm if the original A and/or E horizons were less than 20 cm thick. Throughout most cultivated areas of class 2 erosion, the surface layer consists of a mixture of the original A and/or E horizons and material from below. Some areas may have intricate patterns, ranging from uneroded small areas to severely eroded small areas. Where the original A and/or E horizons were very thick, little or no mixing of underlying material may have taken place.

Class 3. This class consists of soils that have lost, on the average, 75 percent or more of the original A and/or E horizons or of the uppermost 20 cm if the original A and/or E horizons were less than 20 cm thick. In most areas of class 3 erosion, material below the original A and/or E horizons is exposed at the surface in cultivated areas; the plow layer consists entirely or largely of this material.

Even where the original A and/or E horizons were very thick, at least some mixing with underlying material generally took place.

Class 4. This class consists of soils that have lost all of the original A and/or E horizons or the uppermost 20 cm if the original A and/or E horizons were less than 20 cm thick. In addition, Class 4 includes some or all of the deeper horizons throughout most of the area. The original soil can be identified only in small areas. Some areas may be smooth, but most have an intricate pattern of gullies.

Soil Water

This section discusses "the water regime"—schemes for the description of the state of the soil water at a particular time and for the change in soil water state over time. Soil water state is evaluated from water suction, quantity of water, whether the soil water is liquid or frozen, and the occurrence of free water within the soil and on the land surface. Complexity and detail of water regime statements may range widely.

Inundation Classes

Free water may occur above the soil. Inundation is the condition that the soil area is covered by liquid free water. Flooding is temporary inundation by flowing water. If the water is standing, as in a closed depression, the term ponding is used.

Internal Classes

Definitions.—Table below contains water state classes for the description of individual layers or horizons. Only matrix suction is considered in definition of the classes. Osmotic potential is not considered.

For water contents of medium and fine-textured soil materials at suctions less than about 200 kPa, the reference laboratory water retention is for the natural soil fabric. Class limits are expressed both in terms of suction and water content. In order to make field and field office evaluation more practicable, water content pertains to gravimetric quantities and not to volumetric.

The classes are applicable to organic as well as to mineral soil material. The frozen condition is indicated separately by the symbol "f." The symbol indicates the presence of ice; some of the water may not be frozen. If the soil is frozen, the water content or suction pertains to what it would be if not frozen.

Table: *Water state classes*

Class	*Criteria*[a]
Dry (D)	>1500 kPa suction
Very Dry (DV)	<(0.35 x 1500 kPa retention)
Moderately Dry (DM)	0.35 to 0.8 x 1500 kPa retention
Slightly Dry (DS)	0.8 to 1.0 x 1500 kPa retention
Moist (M)	<1500 kPa to >1 or 1/2 kPa[b]
Slightly Moist (MS)	1500 kPa suction to MWR[c]
Moderately Moist (MM)	MWR to UWR[c]
Very Moist (MV)	UWR to 1 or 1/2 kPa[b] suction
Wet <1 kPa or <1/2 kPa[b]	
Nonsatiated (WN)	No free water
Satiated (WA)	Free water present

a. Criteria use both suction and gravimetric water contents as defined by suction.

b. 1/2 kPa only if coarse soil material.c. UWR is the abbreviation for upper water retention, which is the laboratory water retention at 5 kPa for coarse soil material and 10 kPa for other. MWR is the midpoint water retention. It is halfway between the upper water retention and the retention at 1500 kPa.

Three classes and eight subclasses are defined. Classes and subclasses may be combined as desired. Symbols for the combinations currently defined are in table 2. Specificity desired and characteristics of the water desorption curve would determine whether classes or subclasses would be used. Coarse soil material has little water below the 1500 kPa retention, and so subdivisions of dry generally would be less useful.

Dry is separated from *moist* at 1500 kPa suction. *Wet* is separated from moist at the condition where water films are readily apparent. The water suction at the moist-wet boundary is assumed to be about 1/2 kPa for coarse soil materials and 1 kPa for other materials. The formal definition of coarse soil material is given later.

Three subclasses of dry are defined—*very dry, moderately dry*, and *slightly dry*. Very dry cannot be readily distinguished from air dry in the field. The water content extends from ovendry to 0.35 times the water retention at 1500 kPa. The upper limit is roughly 150 percent of the air dry water content. The limit between moderately dry and slightly dry is a water content 0.8 times the retention at 1500 kPa.

The moist class is subdivided into *slightly moist, moderately moist,* and *very moist.* Depending on the kind of soil material, laboratory retention at 5 or 10 kPa suction (method 4B, Soil Survey Laboratory Staff, 1992) determines the *upper water retention.* A suction of 5 kPa is employed for coarse soil material. Otherwise, 10 kPa is used.

To be considered coarse, the soil material that is strongly influenced by volcanic ejecta must be nonmedial and weakly or nonvesicular. If not strongly influenced by volcanic ejecta, it must meet the sandy or sandy-skeletal family particle size criteria and also be coarser than loamy fine sand, have <2 percent organic carbon, and have <5 percent water at 1500 kPa suction. Furthermore, the computed total porosity of the <2 mm fabric must exceed 35 percent.

Very moist has an upper limit at the moist-wet boundary and a lower limit at the upper water retention. Relatedly, *moderately moist* has an upper limit at the upper water retention and a lower limit at the midpoint in gravimetric water content between retention at 1500 kPa and the upper water retention. This lower limit is referred to as the midpoint water retention. *Slightly moist,* in turn, extends from the midpoint water retention to the 1500 kPa retention.

The wet class has *nonsatiated* and *satiated* subclasses distinguished on the basis of absence or presence of free water. Miller and Bresler (1977) defined satiation as the condition from the first appearance of free water through saturation. The nonsatiated wet state may be applicable at zero suction to horizons with low or very low saturated hydraulic conductivity. These horizons may not exhibit free water. Horizons may have parts that are *satiated wet* and other parts, because of low matrix saturated hydraulic conductivity and the absence of conducting macroscopic pores, that are *nonsatiated wet.* Free water develops positive pressure with depth below the top of a wet satiated zone.

A class for saturation (that is, zero air-filled porosity) is not provided because the term suggests that all of the pore space is filled with water. This condition usually cannot be evaluated in the field. Further, if saturation is used for the concept of satiation, then a term is not available to describe known saturation. There is an implication of saturation if the soil material is satiated wet and coarse-textured or otherwise has properties indicative of high or very high saturated hydraulic conductivity throughout the mass. A satiated condition does not necessarily indicate reducing conditions. Air may be present in the water and/or the microbiological activity may be low. The presence of reducing conditions may be inferred from soil colour in some instances and a test may be performed for ferrous iron in solution. The results of the test for ferrous iron should be reported separately from the water-state description.

Evaluation.—*Wet* is indicated by the occurrence of prominent water films on surfaces of sand grains and structural units that cause the soil material to glisten. If free water is absent, the term *nonsatiated wet* is used. If free water is present, the term *satiated wet* is used. The position of the upper field boundary of the satiated wet class, in a formal sense, is the top of the water in an unlined bore hole after equilibrium has been reached. Determination of the thickness of a perched zone of free water requires the installation of lined bore holes or piezometers to several depths across the zone of free water occurrence. Piezometers are tubes placed to the designated depth that are open at both ends, may have a perforated zone at the bottom, but do not permit water entry along most of their length. In the context here, information about the depth of free water and location and thickness of the free water zone would be obtained in the course of soil examination for a range of purposes and does not necessarily require installation of bore holes.

Ideally, evaluation within the moist and dry classes should be based on field instrumentation. Usually, such instrumentation is not available and approximations must be made. Gravimetric water content measurements may be used. To make the conversion from measured water content to suction it is necessary to have information on the gravimetric water retention at different suctions. The water retention at 1500 kPa may be estimated from the field clay percentage evaluation if dispersion of clay is relatively complete for the soils of concern. Commonly, the 1500 kPa retention is roughly 0.4 times the clay percentage. This relationship can be refined considerably as the soil material composition and organization is increasingly specified. Another rule of thumb is that the water content at air-dryness is about 10 percent of the clay percentage, assuming complete dispersion. Model-based curves that relate gravimetric water content and suction are available for many soils.

These curves may be used to determine upper water retention and the midpoint water retention, and to place the soil material in a water state class based on gravimetric water contents. Further such curves would be the basis in many instances for estimation of the water retention at 10 kPa from measurements at 33 kPa. A model-based curve for a medium-textured horizon and the relationship of water-state class limits to water contents determined from the desorption curve. The figure includes the results of a set of tests designed to provide local criteria for field and field office evaluation of water state. These will be discussed subsequently.

Commonly, gravimetric water content information is not available. Visual and tactile observations must suffice for the placement. Separation between moist and wet and the distinction between the two subclasses of wet may be made visually, based on water-film expression and presence of free water. Similarly, the separation between very dry and moderately dry can be made by visual or tactile comparison of the soil material at the field water content and after air drying. The change on air drying should be quite small, if the soil material initially is in the very dry class.

Criteria are more difficult to formulate for soil material that is between the moist/wet and the moderately dry/very dry separations. Four tests follow that may be useful for mineral soils. The three tests that involve tactile examination are performed on soil material that has been manipulated and mixed. This manipulation and mixing may change the tactile qualities from that of weakly altered soil material. The change may be particularly large for dense soil. In the field, this limitation should be kept in mind.

Colour value test. The crushed colour value of the soil for an unspecified water state is compared to the colour value at air dryness and while moderately moist or very moist. This test probably has usefulness only if the full range of colour value from air dry to moderately moist exceeds one unit of colour value. The change in colour value and its interpretation depends on the water desorption characteristics of the soil material. For example, as the water retention at 1500 kPa increases, the difference between the minimum colour value in the dry state and the very moist colour value tends to decrease. Ball test. A quantity of soil is squeezed firmly in the palm of the hand to form a ball about 3 to 4 cm in diameter. This is done in about five squeezes. The sphere should be near the maximum density that can be obtained by squeezing. Preparation of the ball will differ among people. The important point is that the procedure is consistent for an individual.

In one approach, the ball is dropped from progressively increasing heights onto a nonresilient surface. The height in centimetres at which rupture occurs is recorded. Usually heights above 100 cm are not measured. Additionally, the manner of rupture is recorded. If the ball flattens and does not rupture, the term "deforms" is used. If the ball breaks into about five or less units, the term "pieces" is used. Finally, if the number of units exceeds about five, the term "crumbles" is used.

Alternatively, penetration resistance may be used. The penetrometer is inserted in the ball in the same fashion as would be done for soil in place. This alternative is only applicable for medium and fine- textured soil materials at higher water contents because these soil materials are relatively plastic and not subject to cracking.

Rod test. The soil material is rolled between thumb and first finger or on a surface to form a rod 3 mm in diameter or less. This rod must remain intact while being held vertically from an end for recognition as a rod. Minimum length required is 2 cm. If the maximum length that can be formed is 2 to 5 cm, the rod is weak. If the maximum length equals or exceeds 5 cm, the rod is strong.

The rod test has close similarities to the plastic limit test. Plastic limit values exceed the 1500 kPa retention at moderate clay contents and approach but are not commonly lower than the 1500 kPa retention at high clay contents. If a strong rod can be formed, the water content usually exceeds the 1500 kPa retention. The same is probably true for a weak rod. An adjustment is necessary if material of 2 to 0.5 mm is present because the plastic limit is measured on material that passes a number 40 sieve (0.43 mm in diameter).

Ribbon test. The soil material is smeared out between thumb and first finger to form a flattened body about 2 mm of thickness. The minimum length of a coherent unit required for recognition of a ribbon is 2 cm. If the maximum length is 2 to 4 cm, the ribbon is weak. If the maximum length equals or exceeds 4 cm, the ribbon is strong.

To establish criteria based on the foregoing tests it is highly desirable to apply the tests first to soil materials that are known to be at water-state class limits. The approach would parallel that used to maintain quality control of field texture evaluation. The first step to obtain such samples is to establish gravimetric water contents for the class limits. Soil material is prepared at these water contents. A known weight of soil material at a measured, initially higher, water content than the desired final content is placed in a commercial, nylon oven-cooking bag. These bags pass from 1 to 10 grams per hour of water at room temperature, depending on the size, the air temperature, humidity, and movement. Water loss from the bag is continued until the predetermined weight (hence, desired water content) is reached. If long-term storage is desired, the soil is next transferred to glass canning jars. The soil material either may be dried from an initially higher field water content after passing through a number 4 sieve (4.8 mm) or may be air-dried, ground, wetted to above the desired final water content, and then dried. It is preferable to pass the soil through a number 4 sieve (4.8 mm) rather than a number 10 (2 mm). The natural organization is retained to a greater extent. As a result, the calibration sample feels more like it would under field conditions. For the higher sections, consideration should be given to storage of the soil material for a day or two after the water content reduction to improve equilibration.

General relationships of the tests to water state, with the exception of the relationship of the rod test to 1500 kPa retention, have not been formulated and are probably not feasible. The tests may be applied to groupings of soils based on composition, and then locally applicable field criteria can be formulated.

Natural Drainage Classes

Natural drainage class refers to the frequency and duration of wet periods under conditions similar to those under which the soil developed. Alteration of the water regime by man, either through drainage or irrigation, is not a consideration unless the alterations have significantly changed the morphology of the soil. The classes follow:

Excessively drained. Water is removed very rapidly. The occurrence of internal free water commonly is very rare or very deep. The soils are commonly coarse-textured and have very high hydraulic conductivity or are very shallow.

Somewhat excessively drained. Water is removed from the soil rapidly. Internal free water occurrence commonly is very rare or very deep. The soils are commonly coarse-textured and have high saturated hydraulic conductivity or are very shallow.

Well drained. Water is removed from the soil readily but not rapidly. Internal free water occurrence commonly is deep or very deep; annual duration is not specified. Water is available to plants throughout most of the growing season in humid regions. Wetness does not inhibit growth of roots for significant periods during most growing seasons. The soils are mainly free of the deep to redoximorphic features that are related to wetness.

Moderately well drained. Water is removed from the soil somewhat slowly during some periods of the year. Internal free water occurrence commonly is moderately deep and transitory through permanent. The soils are wet for only a short time within the rooting depth during the growing season, but long enough that most mesophytic crops are affected. They commonly have a moderately low or lower saturated hydraulic conductivity in a layer within the upper 1 m, periodically receive high rainfall, or both.

Somewhat poorly drained. Water is removed slowly so that the soil is wet at a shallow depth for significant periods during the growing season. The occurrence of internal free water commonly is shallow to moderately deep and transitory to permanent. Wetness markedly restricts the growth of mesophytic crops, unless artificial drainage is provided. The soils commonly have one or more of the following characteristics: low or very low saturated hydraulic conductivity, a high water table, additional water from seepage, or nearly continuous rainfall.

Poorly drained. Water is removed so slowly that the soil is wet at shallow depths periodically during the growing season or remains wet for long periods. The occurrence of internal free water is shallow or very shallow and common or persistent. Free water is commonly at or near the surface long enough during the growing season so that most mesophytic crops cannot be grown, unless the soil is artificially drained. The soil, however, is not continuously wet directly below plow-depth. Free water at shallow depth is usually present. This water table is commonly the result of low or very low saturated hydraulic conductivity of nearly continuous rainfall, or of a combination of these.

Very poorly drained. Water is removed from the soil so slowly that free water remains at or very near the ground surface during much of the growing season. The occurrence of internal free water is very shallow and persistent or permanent. Unless the soil is artificially drained, most mesophytic crops cannot be grown. The soils are commonly level or depressed and frequently ponded. If rainfall is high or nearly continuous, slope gradients may be greater.

Infiltration

Infiltration is the process of downward water entry into the soil. The values are usually sensitive to near surface conditions as well as to the antecedent water state. Hence, they are subject to significant change with soil use and management and time.

Infiltration stages.—Three stages of infiltration may be recognized—preponded, transient ponded, and steady ponded. *Preponded infiltration* pertains to downward water entry into the soil under conditions that free water is absent on the land surface. The rate of water addition determines the rate of water entry. If rainfall intensity increases twofold, then the infiltration increases twofold. In this stage, surface-connected macropores are relatively ineffective in transporting water downward. No runoff occurs during this stage.

As water addition continues, the point may be reached where free water occurs on the ground surface. This condition is called ponding. The term in this context is less restrictive than its use in inundation. The free water may be restricted to depressions and be absent from the majority of the ground surface. Once ponding has taken place, the control over the infiltration shifts from the rate of water addition to characteristics of the soil. Surface-connected nonmatrix and subsurface-initiated cracks then become effective in transporting water downward.

Infiltration under conditions where free water is present on the ground surface is referred to as *ponded infiltration*. In the initial stages of ponded infiltration, the rate of water entry usually decreases appreciably with time because of the deeper wetting of the soil, which results in a reduced suction gradient, and the closing of cracks and other surface-connected macropores. *Transient ponded infiltration* is the stage at which the ponded infiltration decreases markedly with time. After long continued wetting under ponded conditions, the rate of infiltration becomes steady. This stage is referred to as *steady ponded infiltration*. Surface-connected cracks would be closed, if reversible. The suction gradient would be small and the driving force reduced to near that of the gravitational gradient. Assuming the absence of ice and of zones of free water within moderate depths and that surface or near surface features (crust, for example) do not control infiltration, the minimum saturated hydraulic conductivity within a depth of 1/2 to 1 meter should be a useful predictor of steady ponded infiltration rate.

Minimum Annual Steady Ponded Infiltration.—The steady ponded infiltration rate while the soil is in the wettest state that regularly occurs while not frozen is called the *minimum annual steady ponded infiltration rate*. The quantity is subject to reduction because of the presence of free water at shallow depths if this is a predictable feature of the soil. Allowance for the effect of free water differentiates the quantity from minimum saturated hydraulic conductivity for the upper meter of the soil. The minimum annual steady ponded infiltration rate has application for prediction of runoff at the wettest times of the year when the runoff potential should be the highest.

Soil Temperature

Soil temperature exerts a strong influence on biological activities. It also influences the rates of chemical and physical processes within the soil. When the soil is frozen, biological activities and chemical processes essentially stop. Physical processes that are associated with ice formation are active if unfrozen zones are associated with freezing zones. Below a soil temperature of about 5 °C, growth of roots of most plants is negligible. In areas where soils have permanently frozen layers near the surface, however, even large roots of adapted plants are present immediately above the frozen layer late in the summer. Most plants grow best within a restricted range of soil and air temperature. Knowledge of soil and air temperature is essential in understanding soil-plant relationships. Temperature changes with time, as does the soil-water state. It generally differs from layer to layer at any given time.

Characteristics of Soil Temperature

Heat is both absorbed at and lost from the surface of the soil. Temperature at the surface can change in daily cycles. The soil transmits heat downward when the temperature near the surface is higher than the temperature below and heat upward when the temperature is warmer within the soil than at the surface. Soil temperatures at various depths within the soil follow cycles. The cycles deeper in the soil lag behind those near the surface.

The daily cycles decrease in amplitude as depth increases and are scarcely measurable below 50 cm in most soils. Seasonal cycles are evident to much greater depths if seasonal air temperature differences are pronounced, but the temperature at a depth of 10 m is nearly constant in most soils and is about the same as the mean annual temperature of the soil above.

Soil temperature varies from layer to layer at a given site at a given time; yet, if the average annual temperatures at different depths in the same pedon are compared, they usually do not differ. Mean annual temperature is one of several useful values that describe the temperature regime of a soil. The seasonal fluctuation of soil temperature is a characteristic of a soil. Soil temperature fluctuates little seasonally near the equator; it fluctuates widely as seasons change in the middle and high latitudes. Mean seasonal temperatures can be used to characterize soil temperature. Seasonal temperature differences decrease and the seasonal cycles lag progressively as depth increases.

For soils that freeze in winter, soil temperature is influenced by the release of heat when water changes from the liquid to the solid form. This releases about 80 calories per gram of water. The heat must be dissipated before the water in soil freezes. The rate of thaw of frozen soils is slower, because heat is required to warm the soil in order to melt the ice. In areas of heavy snowfall, the snow provides an insulating blanket and soils do not freeze as deeply or may not freeze at all.

Many factors influence soil temperature. They include amount, intensity, and distribution of precipitation; daily and monthly fluctuations in air temperature; insolation; kinds, amounts, and persistence of vegetation; duration of moisture states and snow cover; kinds of organic deposits; soil colour; aspect and gradient of slope; elevation; and ground water. All of these factors may be described in a soil survey if they are significant.

Estimating Soil Temperature

Mean annual soil temperature in temperate, humid, continental climates can be approximated by adding 1 °C to the mean annual air temperature reported by standard meteorological stations at locations representative of the soil to be characterized. The mean annual soil temperature at a given place can be estimated more reliably by a single reading at a depth of 10 m.

If water in wells is at depths between 10 and 20 m, the temperature of the water usually gives a close estimate of mean annual soil temperature. Mean annual soil temperature can also be estimated closely from the average of four readings at about 50 cm or greater depth, equally spaced throughout the year.

The mean soil temperature for summer can be estimated by averaging three measurements taken at a constant depth between 50 cm and 1 m on the 15th of each of the three months of the season. Similar methods may be used to estimate soil temperature for other seasons. These methods give values subject to minor variation caused by differences in vegetation (particularly density of canopy), ground water, snow, aspect, rain, unusual weather conditions, and other factors. Tests for nearly level, freely drained soils, both grass-covered and cultivated, produce comparable values. Over the usual period of a soil survey, systematic studies can be made to establish temperature relationships in the survey area.

Soils vary widely in the degree to which horizons are expressed. Relatively fresh geologic formations, such as fresh alluvium, sand dunes, or blankets of volcanic ash, may have no recognizable genetic horizons, although they may have distinct layers that reflect different modes of deposition. As soil formation proceeds, horizons may be detected in their early stages only by very careful examination. As age increases, horizons generally are more easily identified in the field. Only one or two different horizons may be readily apparent in some very old, deeply weathered soils in tropical areas where annual precipitation is high.

Layers of different kinds are identified by symbols. Designations are provided for layers that have been changed by soil formation and for those that have not. Each horizon designation indicates either that the original material has been changed in certain ways or that there has been little or no change. The designation is assigned after comparison of the observed properties of the layer with properties inferred for the material before it was affected by soil formation. The processes that have caused the change need not be known; properties of soils relative to those of an estimated parent material are the criteria for judgment. The parent material inferred for the horizon in question, not the material below the solum, is used as the basis of comparison. The inferred parent material commonly is very similar to, or the same as, the soil material below the solum.

Designations show the investigator's interpretations of genetic relationships among the layers within a soil. Layers need not be identified by symbols for a good description; yet, the usefulness of soil descriptions is greatly enhanced by the proper use of designations.

Designations are not substitutes for descriptions. If both designations and adequate descriptions of a soil are provided, the reader has the interpretation made by the person who described the soil and also the evidence on which the interpretation was based.

Genetic horizons are not equivalent to the diagnostic horizons of Soil Taxonomy. Designations of genetic horizons express a qualitative judgment about the kind of changes that are believed to have taken place. Diagnostic horizons are quantitatively defined features used to differentiate among taxa. Changes implied by genetic horizon designations may not be large enough to justify recognition of diagnostic criteria. For example, a designation of Bt does not always indicate an argillic horizon. Furthermore, the diagnostic horizons may not be coextensive with genetic horizons.

Three kinds of symbols are used in various combinations to designate horizons and layers. These are capital letters, lower case letters, and Arabic numerals. Capital letters are used to designate the master horizons and layers; lower case letters are used as suffixes to indicate specific characteristics of master horizons and layers; and Arabic numerals are used both as suffixes to indicate vertical subdivisions within a horizon or layer and as prefixes to indicate discontinuities.

Chapter 8

Soil Analysis

Soil pH

The soil pH measures active soil acidity or alkalinity. A pH of 7.0 is neutral. Values lower than 7.0 are acid; values higher are alkaline. Usually the most desirable pH range for mineral soils is 6.0 to 7.0 and for organic soils 5.0 to 5.5. The soil pH is the value that should be maintained in the pH range most desirable for the crop to be grown.

Buffer pH

This is an index value used for determining the amount of lime to apply on acid soils to bring the pH to the desired pH for the crop to be grown. The lower the buffer pH reading the higher the lime requirement.

Phosphorus

The phosphorus test measures that phosphorus that should be available to the plant. The optimum level will vary with crop, yield and soil conditions, but for most field crops a medium to optimum rating is adequate. For soils with pH above 7.3 the sodium bicarbonate test will determine the available P.

Potassium

This test measures available potassium. The optimum level will vary with crop, yield, soil type, soil physical condition, and other soil related factors. Generally higher levels of potassium are needed on soils high in clay and organic matter versus soils, which are sandy and low in organic matter.

Optimum levels for light-coloured, coarse-textured soils may range from 90 to 125 ppm (180 to 250 lbs/ac). On dark-coloured heavy-textured soils levels ranging from 125 to 200 ppm (250 to 400 lbs/ac) may be required.

Calcium

Primarily soil type, drainage, liming and cropping practices affect the levels of calcium found in the soil.

Calcium is closely related to soil pH. Calcium deficiencies are rare when soil pH is adequate. The level for calcium will vary with soil type, but optimum ranges are normally in the 65% to 75% cation saturation range.

Magnesium

The same factors, which affect calcium levels in the soil, also influence magnesium levels except magnesium deficiencies are more common. Adequate magnesium levels range from 30 to 70 ppm (60 to 140 lbs/ac). The cation saturation for magnesium should be 10 to 15%.

Sulphur

The soil test measures sulfate-sulfur. This is a readily available form preferred by most plants. Soil test levels should be maintained in the optimum range.

It's important that other soil factors, including organic matter content, soil texture and drainage be taken into consideration when interpreting sulfur soil test and predicting crop response.

Boron

The readily soluble boron is extracted from the soil. Boron will most likely be deficient in sandy soils, low in organic matter with adequate rainfall. Soil pH, organic matter level and texture should be considered in interpreting the boron test, as well as the crop to be grown.

Copper

Copper is most likely to be deficient on low organic matter sandy soils, or organic soils. The crop to be grown, soil texture, and organic matter should be considered when interpreting copper tests. A rating of medium to optimum should be maintained.

Iron

Soil pH is a very important factor in interpreting iron tests. In addition, crops vary a great deal in sensitivity to iron deficiency. Normally a medium level would be adequate for most soils. If iron is needed it would be best applied foliar.

Manganese Soil

pH is especially important in interpreting manganese test levels. In addition, soil organic matter, crop and yield levels must be considered. Manganese will work best if applied foliar or banded in the soil.

Zinc

Other factors, which should be considered in interpreting the zinc test, include available phosphorus, pH, and crop and yield level. For crops that have a good response to zinc, the soil test level should be optimum.

Sodium

Sodium is not an essential plant nutrient but is usually considered in light of its effect on the physical condition of the soil. Soils high in exchangeable sodium may cause adverse physical and chemical conditions to develop in the soil. These conditions may prevent the growth of plants. Reclamation of these soils involves the replacement of the exchangeable sodium by calcium and the removal by leaching.

Soluble Salts

Excessive concentration of various salts may develop in soils. This may be a natural occurrence or it may result from irrigation, excessive fertilization or contamination from various chemicals or industrial wastes. One effect of high soil salt concentration is to produce water stress in a crop to where plants may wilt or even die. The effect of salinity is negligible if the reading is less than 1.0 mmhos/cm. Readings greater than 1.0 mmhos/cm may affect salt sensitive plants and readings greater than 2.0 mmhos/cm may require the planting of salt tolerant plants.

Organic Matter and ENR (Estimated Nitrogen Release)

Percent organic matter is a measurement of the amount of plant and animal residue in the soil. The colour of the soil is usually closely related to its organic matter content, with darker soils being higher in organic matter. The organic matter serves as a reserve for many essential nutrients, especially nitrogen. During the growing season, a part of this reserve nitrogen is made available to the plant through bacterial activity. The ENR is an estimate of the amount of nitrogen (lbs/acre) that will be released over the season. In addition to organic matter level, this figure may be influenced by seasonal variation in weather conditions as well as soil physical conditions.

NO_3-N (Nitrate Nitrogen)

Nitrate nitrogen is a measure of the nitrogen available to the plant in nitrate form. In high rainfall areas, sandy soil types and areas with warm winters, this measurement may be of limited value except at planting or side dress time. In the areas with lower rainfall, the nitrate test may be very beneficial.

Cation Exchange Capacity (CEC)

Cation exchange capacity measures the soil's ability to hold nutrients such as calcium, magnesium, and potassium, as well as other positively charged ions such as sodium and hydrogen. The CEC of a soil is dependent upon the amounts and types of clay minerals and organic matter present. The common expression for CEC is in terms of milliequivalents per 100 grams (meq/100g) of soil.

The CEC of soil can range from less than 5 to 35 meq/100g for agricultural type soils. Soils with high CEC will generally have higher levels of clay and organic matter. For example, one would expect soil with a silty clay loam texture to have a considerably higher CEC than a sandy loam soil. Although high CEC soils can hold more nutrients, it doesn't necessarily mean that they are more productive. Much depends on good soil management.

Cation Saturation

Cation saturation refers to the proportion of the CEC occupied by a given cation (an ion with a positive charge such as calcium, magnesium or potassium). The percentage saturation for each of the cations will usually be within the following ranges:

Calcium: 40 to 80 percent

Magnesium: 10 to 40 percent

Potassium: 1 to 5 percent

Soil Testing

The farmers find it extremely difficult to know the proper type of fertilizer, which would match his soil. In using a fertilizer he must take into account the requirement of his crops and the characteristics of the soil. The basic objective of the soil-testing programme is to give farmers a service leading to better and more economic use of fertilizers and better soil management practices for increasing agricultural production. High crop yields cannot be obtained without applying sufficient fertilizers to overcome existing deficiencies.

Efficient use of fertilizers is a major factor in any programme designed to bring about an economic increase in agricultural production. The farmers involved in such a programme will have to use increasing quantities of fertilizers to achieve the desired yield levels. However the amounts and kinds of fertilizers required for the same crop vary from soil to soil, even field to field on the same soil.

The use of fertilizers without first testing the soil is like taking medicine without first consulting a physician to find out what is needed. It is observed that the fertilizers increase yields and the farmers are aware of this. But are they applying right quantities of the right kind of fertilizers at the right time at the right place to ensure maximum profit?

Without a fertilizer recommendation based upon a soil test, a farmer may be applying too much of a little needed plant food element and too little of another element which is actually the principal factor limiting plant growth. This not only means an uneconomical use of fertilizers, but in some cases crop yields actually may be reduced because of use of the wrong kinds or amounts, or improper use of fertilizers. A fertilizers recommendation from a soil testing laboratory is based on carefully conducted soil analyses and the results of up-to-date agronomic research on the crop, and it therefore is most scientific information available for fertilizing that crop in that field.

Each recommendation based on a soil test takes into account the values obtained by these accurate analysis, the research work so far conducted on the crop in the particular soil areas, and the management practices of the concerned farmer.

The soil test with the resulting fertilizer recommendation is therefore the actual connecting link between agronomic research and its practical application to the farmers' fields. However, soil testing is not an end in itself. It is a means to an end. A farmer who follows only the soil test recommendations is not assured of a good crop. Good crop yields are the result of the application also of other good management practices, such as proper tillage, efficient water management, good seed, and adequate plant protection measures. Soil testing is essential and is the first step in obtaining high yields and maximum returns from the money invested in fertilizers.

How to Collect a Soil Sample

1. Sample each field separately. However, where the areas within a field differ distinctly in crop growth, appearance of the soils, or in elevation, or are known to have been cropped or fertilized and manured differently, divided the filed and sample each area separately.

2. Take a composite sample from each area. Scrape away surface litter, then take a small sample from the surface to plough depth from a number of spots in the field (10 to 15 per acre). Collect these samples in a clean bucket or some such wide container.
3. Where crops have been planted in lines (rows), sample between the lines.
4. Do not sample unusual area. Avoid areas recently fertilized, old bunds, marshy spots, near tress, compost piles, other non-representative locations.
5. Take a uniform thick sample from the surface to plough depth. If a spade or a trowel is used, dig a v-shaped hole, then cut out a uniform thick slice of soil from bottom to op of the exposed soil face, collect the sample on the baled or in your hand and place it in the bucket.
6. Pour the soil from the bucket on a piece of clean cloth or paper and mix thoroughly, discard, by quartering, all but 1 to 2 lbs. of soil. Quarterly may be done by mixing sample well, dividing it into four equal parts, then rejecting two opposite quarters, mixing the remaining two portions, again dividing into four parts and rejecting two opposite quarters, and so on. The sample should be dried in the shade for an hour or two before it goes into the cloth bag container.
7. Each cloth bag should be large enough to hold a pound or two of soil, and should be properly marked to identify the sample.
8. Fill out the soil sample information sheet for each sample. These forms may be sent separately to the laboratory or enclosed with the soil sample.
9. Address the samples to the Soil Chemist, Soil Testing Laboratory, Goal Ghar, Port Blair.
10. Keep a record of the areas sampled and a simple sketch map for reference when you get the soil test and fertilizers recommendation report from the soil testing laboratory.

Role of the Extension Service in Soil Testing

The actual analysis of the sample and the making out of fertilizer recommendation is only part of the soil testing service. To a large measure, the efficiency of this service depends upon the care and effort put froth by extension workers and farmers in the collection and dispatch of samples to the laboratory. Its effectiveness also depends upon the proper follow-through of the fertilizer recommendations, including the establishment of result demonstrations on farmer's fields to induce the farmers to follow the fertilizer recommendations. In this work the staff of the extension service play the most important role, since they are the people directly in contact with the farmers or this reason, the soil chemist in charge of the laboratory must give periodic and through training to the extension staff on these subjects.

Collection of Samples

A useful soil testing service starts with the collection of representative soil samples. A fertilizer recommendation made after analyzing the soil can only as good as the sample on which it is based. Actually the one to ten grams of soil used for each chemical analysis should represent as accurately as possible the entire surface six inches of soil, weighing about 2 million pounds per acre. The importance of taking a representative composite sample is, therefore, self-evident. One field can be treated as a single sampling unit only if it is relatively uniform and does not exceed approximately five acres. Variations in slope, colour, texture, management, and cropping pattern should be taken into account and separate composite soil sample adequately representing the field, small portions of surface soil should be collected to depth of six inches from at least ten well-distributed spots in the field, mixed well, and about ½ kg of representative sample sent to laboratory. Proper sampling tools are essential for collection of good soil samples. For a soft. Moist soil, the soil tube, phowda (spade), or khurpi (trowel) are usually quite satisfactory.

For harder soils, a screw type auger, or an adze might be more convenient. Post hole augers are convenient for sampling excessively wet areas like paddy fields An extension worker whose duties include collection of soil samples should be supplied with at least a few of these tools, and also a plastic bucket. The phowda, khurpi and adze are very common implements available in most hardware shops and so there should be no difficulty in procuring these implements.

The farmers should be given help in filling out the soil sample information sheet with an ex-plantation of any items not understood. It should be remembered that the information sheet is very vital part of procedures that go to make a good soil test recommendation. This sheet must supply all of the background information that, in combination with the results of the analysis, makes possible an accurate fertilizer recommendation for a certain crop, for that particular field.

Factors such as crop variety, slope of land, irrigation and drainage facilities, and pervious cropping seasons affect the amounts of fertilizer to be applied to particular crop. Any peculiarities noted in the soil or in the vigor or the crop would be very valuable information on the soil sample information sheet as a basis for making an adequate fertilizer recommendation. In the absence of this information, the soil chemist must base his recommendation upon the soil test values alone, and more often than not the farmer will receive an adequate fertilizer recommendation.

Soil Analysis: A key to Soil Nutrient Mangement

High yields of top-quality crops require an abundant supply of 16 essential nutrient elements. In addition to providing a place for crops to grow, soil is the source for most of the essential nutrients required by the crop. Our soil resource can be compared to a bank where continued withdrawal without repayment cannot continue indefinitely. As nutrients are removed by one crop and not replaced for subsequent crop production, yields will decrease accordingly. Accurate accounting of nutrient removal and replacement, crop production statistics, and soil analysis results will help the producer manage fertilizer applications.

A soil analysis is used to determine the level of nutrients found in a soil sample. As such, it can only be as accurate as the sample taken in a particular field. The results of a soil analysis provide the agricultural producer with an estimate of the amount of fertilizer nutrients needed to supplement those in the soil. Applying the appropriate type and amount of needed fertilizer will give the agricultural a more reasonable chance to obtain the desired crop yield.

Objectives of Soil Analysis

- To provide an index of nutrient availability or supply in a given soil. The soil extract is designed to evaluate a portion of the nutrients from the same "pool" used by the plant.
- To predict the probability of obtaining a profitable response to fertilizer application. Low analysis soils may not always respond to fertilizer applications due to other limiting factors. However, the probability of a response is greater than on a high analysis soil.
- To provide a basis for fertilizer recommendations for a given crop.
- To evaluate the fertility status of the soil and plan a nutrient management program.

Chemical analysis of plant composition indicates chemicals or elements present in a crop at maturity or when it is harvested. For example, 1,250 lb of lint cotton contains approximately 125 lb of nitrogen (N), 20 lb of phosphorus (P), and 75 lb of potassium (K).

The essential question in fertilization is, "How much nutrient must be added to the soil as fertilizer for a given amount to be taken up by the growing plant?" The crop utilizes only a portion of the available nutrients in the soil. This means that more nutrients must be present than are removed by the crop. The amount added varies according to the level already present in the soil and the crop's need for the nutrient involved. The soil analysis is the starting point, since it measures the level or content presently in the soil.

The soil analysis along with the information provided in the information sheet, is interpreted and reported in terms of the nutrients needed to supplement those in the soil. With this information, producers can add sufficient nutrients for the correct balance to obtain high yields.

Limiting Factors

Crop yields are determined by a variety of factors including crop variety selection, available moisture, soil fertility, crop adaptation to the area, and the presence of diseases, insects, and weeds. The soil analysis and its interpretation deal only with the fertility level (plant nutrients) of the soil. Recommended fertilizer will provide sufficient nutrients for the best possible yields. Other factors of production or management may still cause low yields, even though nutrients are adequate.

Carryover

If yields are only partial in relation to a large amount of fertilizer applied, many of the nutrients are carried over for use by the next crop. It is this carryover, or residual effect, from one year to the next that makes heavy fertilizer applications practical in the face of other limits to yield.

Yields to Expect

A certain fertilizer application cannot be expected to produce a specific yield such as two bales of cotton or nine tons of hay. It is more realistic to assume that a balanced fertilizer program assures that nutrients are not the limiting factor in yields obtained. Research has shown that producers who use a balanced fertilizer program obtain consistently better yields than those who don't.

The Soil Analysis Report

After the soil is analyzed, fertility recommendations are made based on amounts of actual nutrients in the soil, not on the amount of any particular fertilizer or mixture. For example, if 100 lb of N were recommended, that amount could be supplied by approximately 300 lb of ammonium nitrate (33%N), 220 lb of urea (45%N), or 120 lb of anhydrous ammonia (82%N). Likewise, a recommendation of 60 lb of P205 per acre could be added as 133 lb of 45% triple superphosphate.

Fertilizer Labeling

Nitrogen is expressed on the elemental basis as "total nitrogen" (N). Phosphorus is expressed on the oxide basis as "available phosphoric acid" (P205). Potassium is expressed as "soluble potash" or potassium oxide (K20).

In reality, there is no P205 or K20 in fertilizers. Phosphorus exists most commonly as monocalcium phosphate, but also occurs as other calcium or ammonium phosphates. Potassium is ordinarily in the form of potassium chloride or sulfate. Furthermore, P205 and K20 are not absorbed by plants. Plant roots absorb most of their phosphorus in the form of orthophosphate ions, H2P04-, and most of their potassium as potassium ions, K+. For these reasons, the elemental expression (N-P-K) is used in all of the recent research publications. Conversions from one form of P and K to another can be made using the following formulas.

%P=%P205 x0.437 %K=%K20x0.826

%P205=%Px2.29 %K20=%Kx1.21

Interpretation of the Soil Analysis Report

The soil analysis report contains two parts: characterization and fertility status of the soil, and fertility recommendations. Soil characterization (pH, texture, percent exchangeable sodium, percent organic matter, and salinity expressed as electrical conductivity) is explained in the report. The fertility status is reported as nutrients available to the plant. The second part, fertility recommendation, contains the suggested amounts of fertilizer to apply. These amounts are based on the crop requirements, management practices affecting the crop (as shown in the information sheet), the present fertility level of the soil, and the yield goal desired by the producer. Special notification is given if the tests indicate that a salt or sodium hazard exists or if the information provided shows any other specific problems.

Soil amendments or treatments to reduce a sodium or salt hazard will be recommended if requested. In general, application of gypsum is suggested for reducing a sodium hazard, and leaching is recommended in most cases to lower salt content in the soil. Gypsum or leaching requirements are calculated and reported if requested.

Where to Get Soil Analyzed

There are many soil testing laboratories in New Mexico, Texas, Colorado, and Arizona. Basic soil testing packages vary in price and number of analyses. Many labs are participating in the Western Region Soil Testing Proficiency Program. Program participants share identical soils and compare results quarterly. This process assures the clients that the lab is striving for consistency and accuracy in lab analyses. Recommendations will undoubtedly vary from lab to lab. Often the best recommendation will come from the local Extension service. The choice of labs is at the client's discretion but should be based on report readability, result accuracy, turn-around time, and cost factors. New Mexico specialists can assist with many questions regarding plant health. Remember, a soil analysis is only as good as the soil sample taken.

Soil Analysis: Key to Nutrient Management Planning

Soil provides a reservoir of nutrients required by crops and also therefore for animals but not necessarily at optimum levels of immediate availability to plants. The purpose of soil analysis is to assess the adequacy, surplus or deficiency of available nutrients for crop growth and to monitor change brought about by farming practices. This information is needed for optimum production, to avoid transferring undesirable levels of some nutrients into the environment and to ensure a suitable nutrient content in crop products.

Farm assurance schemes, buyer's protocols and codes of practice are increasingly demanding more accurate fertiliser recommendations which must depend on the nutrient-supplying capacity of the soil. Regular soil analysis should be undertaken as a vital part of good management practice.

What is Measured

pH, Phosphate, Potash & Magnesium

The "standard soil analysis package" measures soil acidity (pH) and estimates the plant-availableconcentrations of the major nutrients in the soil-phosphorus (P), potassium (K) and magnesium (Mg).

Soil acidity is measured as pH (the concentration of H^+ ions) the scale running from pH 1 (very acid), through pH 7 (neutral) to pH 14 (very alkaline). The normal soil range is pH 4.5-8.5.

The total P and K content of a soil can be measured exactly but has little relevance to crop yield because only a relatively small proportion of the total P and K in soils is available to the plant. Soil analysis in the laboratory therefore uses chemical extractants to provide an estimate of the nutrient which would be available under field conditions. The results provide the best practical guide for determining P, K and Mg in the readily plant-available pool shown in the diagram. The methods of soil analysis used in the UK have been developed over many years and have been correlated to crop response on a wide range of soils in numerous field experiments. Other measurements of nutrient content or ratios may be made on soils, but unless there is dependable correlation with yields, they are of limited practical value.

Calcium (Ca) and sodium (Na) can also be measured, but these are rarely determined.

Basal Nutrients in Soil

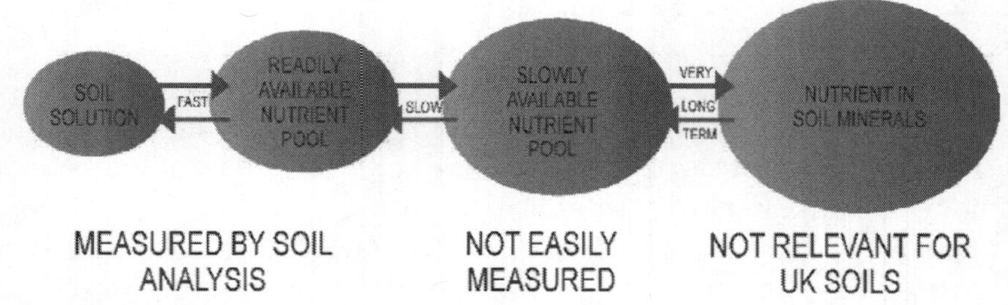

Trace Elements

Of the elements required in small amounts by crops-the 'trace' elements or 'micronutrients'-some can be measured effectively in soil, namely boron (B), chloride (Cl), copper (Cu), molybdenum (Mo), cobalt (Co) and zinc (Zn).

Other elements which are not effectively assessed by soil measurements and need to be measured in herbage or in the animal include manganese (Mn), iodine (I) and selenium (Se).

Heavy Metals

In the UK there are statutory soil limits for 7 potentially toxic elements (PTEs) to ensure compliance with EU legislation when sewage sludge is to be applied. These are the 'heavy metals' cadmium (Cd), chromium (Cr), copper (Cu), lead (Pb), mercury (Hg), nickel (Ni) and zinc (Zn). The total (acid extractable) content of each of these in soil is analysed. The regulations also link the level of these elements to pH which affects their availability to plants. The limits for PTEs are defined in the Sludge (Use in Agriculture) Regulations 1989 and given in the Defra Soil Code. The Code advises the same limits be used to guide the application of other metal containing organic manures and wastes applied to land.

Nitrogen and Sulphur

Measuring the availability of soil nitrogen (N) and sulphur (S) can be useful but is more complex than for other nutrients; these are not covered in this leaflet.

Factors Affecting Soil Analysis

Nutrient values can vary as a result of a number of factors which are discussed below. Further study is needed to define the scale and cause of such variation.

Spatial Variation

Variation can exist over very short distances (less than 1 metre) and the number of cores that are needed to ensure that sampling is representative of an area will vary according to the scale of this variation. When sampling to obtain an average value for a field this is not so important because all the cores taken are bulked into a single sample; for a typically uniform field 25 cores are sufficient. However, local variation is important when grid sampling because each grid point is viewed as a separate sample.

Normally, because the extent of the variation is unknown, at least 16 cores need to be taken on 1 metre grid-spacing around each grid point to provide a single bulked sample for analysis. If it is planned to produce a 'map' of the nutrient variability within the field, then the grid points should be not more than 50 metres apart. Where considerable differences occur within a field, a single sample can be misleading because the averaging will disguise different treatment requirements. If different areas are known to vary they should be sampled separately. Knowledge of field amalgamations, soil type changes observed when ploughing or cultivating, visual crop growth difference across the field or any other specific information on field variation should be used to ensure that the area is sampled to best effect.

Temporal Variation

Variations in nutrient values have been observed at different times of the year and there is evidence to suggest that soil P, K and Mg values will be higher in the early spring than in the autumn as a result of chemical weathering over winter, biological activity and lack of uptake by growing crops

Moisture

If soils are dry at the time of sampling the analytical results can be affected and may appear a little lower for pH, P and possibly K.

Movement and uptake of all nutrients will of course be restricted in very dry soils but this is a transient problem and does not reflect the normal availability of nutrients in the soil. At present it is not possible to quantify this effect in order to improve interpretation of analytical results.

Crop Removal

During periods of rapid growth crop uptake, especially of potassium, can be large and may deplete available soil nutrient levels for a short period until the nutrient status returns to equilibrium. This could affect results for some soils if samples are taken at such times.

Nutrients are returned to the soil in crop residues but will not be determined by analysis until the plant material is broken down. It has been suggested that sampling close to harvest may result in an underestimation of true soil nutrient status.

Sampling Depth

Frequently there may be a gradient in nutrient level down the soil profile, usually declining with depth, reflecting the accumulation of nutrients in the plough layer. This is accentuated in minimal cultivation systems where phosphate and potash may be concentrated in the top 2 or 3 inches (5 to 8 cm). With continuous direct drilling there may be a large difference between the top 2 inches (5 cm) and 3 to 6 inch (8 to 15 cm) zones. For permanent grass where there is no soil disturbance, consistency of sampling depth of 3 inches (8 cm) is particularly important.

Previous Manuring

Applications of fertilisers and manures obviously have a major impact on measurement of soil nutrients. It is difficult to lay down rules as to how soon sampling should be undertaken after application. General guidelines are as follows:

8 weeks after P, K, Mg fertiliser application.12 weeks after slurry or manure application.12 months after lime spreading.

Ploughing and cultivating help distribute nutrients from fertilisers and manures, and lime, throughout the depth of cultivation, but this takes time.

Sampling

Soil analysis data are only as good as the sample taken. A sample normally comprises around 1 kg (2 lbs) of soil which is taken to represent an entire area or field, which contains around 2,000 t of soil per hectare to a plough depth of 8 inches (20 cm). It is therefore imperative to obtain as representative a sample as possible or the results will not reflect the nutrient status accurately.

Rules of Sampling

1. Use of a suitable tool (cone auger, screw auger, corer etc) which facilitates and encourages the taking of more rather than fewer cores, of a uniform size and down to the full depth of sampling.
2. Use equipment and packaging that will not contaminate the sample. Galvanised sampling tools are unacceptable for trace element analysis.
3. Label samples clearly.
4. Sample to a consistent depth. Normal depth is 6 inches (15 cm) for arable soils, 3 inches (7.5 cm) for grassland.
5. Divide the field into areas which are as uniform as possible in soil type, past cropping, and manuring history and sample separately. Small areas of different soil e.g. wet, chalky, shallow, stony etc. should be excluded.
6. Avoid headlands, gateways, trees, mole hills, dung/urine patches, water troughs, areas where lime or manure has been dumped, old hedgerows/middens/ponds or any other irregular feature.
7. Discard stone and plant debris.
8. Take at least 25 cores from each area to be sampled and put them together to form a single representative sample. The numbers of cores should not be restricted simply because the container is full! Thoroughly mix all cores and take a sub-sample from this for despatch to the laboratory-this must be done carefully.
9. Ensure the sample represents the whole area. Sample on a W pattern over the field; for a regular shaped field this means 7 cores per leg of the "W".
10. In the case of grid sampling up to 16 cores are needed at each sampling point to obtain a representative sample. These should be taken in a regular pattern about 1 metre apart around the grid point. Grid points should be evenly spaced over the field and should not be more than 50 metres apart if a map is to be produced.
11. Sample at the same point in the rotation, before the crop which is most demanding or responsive to P and K. In

descending order of importance these are: horticultural crops, vegetables, roots, pulses, spring sown combinable crops, winter sown combinable crops. For pH it is preferable to sample 12 months before a sensitive crop such as sugar beet or barley.

12 Sample at the same time of the year.

13 Avoid sampling under extremes of soil conditions e.g. waterlogged or very dry soil.

14 Do not sample within 8 weeks of fertilising, or within 12 weeks of manure or slurry application, for P, K and Mg analysis or sooner than 12 months after liming for pH analysis.

15 Maintain records and use the analytical results to develop nutrient management plans.

16 Where sampling to diagnose a crop problem take multi-cored samples (at least 16 cores per bulked sample) from areas of poor growth and separate samples from normal (good) areas. The relative values between good and poor will be more informative than the actual values of the problem area.

Frequency of Sampling

Soil nutrient levels do not alter markedly over short periods of time unless major factors of supply or demand are introduced. There is therefore no point in incurring the additional cost of sampling more frequently than necessary, except perhaps as a check to ensure that earlier sampling has been carried out properly. Additional sampling may be justified when there is a major change to husbandry practice-for instance alteration of cropping or manure policy. General guidelines for sampling frequency are as follows:-

	Sample every
Permanent grass	7 years
Intensively used grassland	3-4 years
General arable cropping	4-5 years
Arable/grass systems	4-5 years
Field vegetables & horticulture	2-3 years

Analysis

There are a number of reasons why different results may be obtained for identical samples analysed by different laboratories. The most obvious reason is that different extractants may have been used.

Variations in results can also occur if identical protocols are not followed-techniques of soil drying, grinding and sieving, reagent and equipment temperature, extractant concentrations, extraction shaking, stirring and filtration, and extraction time can all affect the result.

Results are usually reported in mg/l but where a laboratory measures soil by weight instead of volume, the units will be as mg/kg. This can give rise to differences on some soils-particularly those with high levels of organic matter. If different laboratories are used for analyses these factors must be taken into account.

Interpretation of Soil pH

In England & Wales target pH (measured in water) is 6.5 for arable crops and 6.0 for grassland. Optimum range in Scotland where Scottish Agricultural College (SAC) measure pH in calcium chloride is 6.0-6.5 for arable crops and 5.7-6.2 for grassland.

In-field measurements using pH indicator on some soils where free chalk or lime particles exist may give lower values than laboratory results for the same field. This is because grinding the soil for laboratory analysis pulverises any chalk/lime particles and the pH as measured is increased. Acidity below pH 6.0 will reduce the availability of nutrients, especially P.

Soil analysis may overstate the availability of K on some alkaline soils where pH is over 7.0 and responses to added potash may occur at higher soil K levels than would be expected. Availability of trace elements is radically affected by pH and the need for trace elements should only be assessed after any required amendment of acidity has been undertaken and has had time to take effect.

Interpretation of Soil P, K & Mg

Soil analysis provides an estimate of available P, K and Mg concentrations in soil to sampling depth-in practice this is equivalent to plough or cultivation depth because of the distribution of nutrients when the land is worked. Response experiments with different crop groups have provided the relationship between crop yield and soil nutrient concentration. Normally, yields increase with increasing nutrient concentration to a maximum, beyond which there is no further benefit from additional nutrient. Below this value, which will vary with crop species, there is a yield penalty. Whilst soil analysis is not a precise guide, the lower the value the greater the risk of poor performance. To aid interpretation of the different concentrations of individual nutrients, Index or descriptive scales are used. These scales provide a general indication of the likely crop response and therefore a guide to the need for additional nutrient supplementation, as shown in the table.

Crop response and soil analysis

		Yield response to added nutrient by	
Defra Index	***SAC description***	***vegetable crops***	***arable crops & grass***
0	Very low	highly likely	highly likely
1	Low	highly likely	probable
2	Moderate	likely	unlikely
3	High	possible	nil
4	High	unlikely	nil
5	High	nil	nil

Soil P, K & Mg Concentrations (mg/l) and Defra Index Scale

Note that the index is split in half for potassium only and described as 2- (or lower index 2) and 2+ (or upper index 2). In the past, index 2 was not divided in half for potassium but some soil reports used + and- signs to denote the extreme top and bottom 10% of each band; laboratories should no longer be using this convention.

Besides providing a basis to decide fertiliser quantities, soil analysis should also be used to monitor changes in fertility especially where there are uncertainties in the amounts of nutrient removed (e.g. with forage crops) and in the amounts of nutrients applied (e.g. with manures and slurries). For this purpose it is desirable to use the mg/l values not the index. However differences of less than 5 mg/l Olsen P, 25 mg/l K and 10 mg/l Mg should be ignored unless part of a sustained trend. Where accurate nutrient balance information is used in conjunction with regular soil analysis, it is important to recognise the possibilities of variation as discussed above.

Soil P, K and Mg Concentrations (mg/l) and SAC Descriptive Scale

The Scottish Agricultural College laboratory uses different extractants to those used in England and Wales and a descriptive rather than a numeric scale.

Description	***Phosphorus***	***Potassium***	***Magnesium***
	Modified Morgans extraction		
Very low	0-10	0-40	0-20
Low	10-25	40-75	20-60
Moderate	26-75	76-200	61-200
High	76-200	201-1000	201-1000
Excessively High	201-	1001-	1001-

Relationship between Defra and SAC Scales

Defra Index	SAC description
0	Very low
1	Low
2	Moderate
3-7	High
8-9	Excessively high

Principles of P, K & Mg Manuring

The principle of manuring is to maintain plant-available soil nutrient levels within a target range depending upon crop rotation and soil type, by replacing nutrients removed.

Soil analysis shows the status of soil nutrients relative to the target values and allows changes as a result of husbandry to be monitored. Where soils are below the target level, nutrient applications should provide more than is removed by the crop to ensure full yield response and to improve nutrient levels.

Nutrient applications for soils above the target range may be reduced or omitted until the soil reserve approaches the target value. Additional nutrients may be applied before very responsive crops such as potatoes with the surplus balance being allowed for in subsequent less responsive crops. The overall nutrient balance in the rotation should be estimated and checked by regular soil analysis.

Sandy Soils

On true sand textured soils (not loamy sands or sandy loams) it is not practical to attempt to maintain soil K at 2- and the target K level should be adjusted to 100-120 mg/l (index 1+).

Chalk and Limestone Soils

At very high pH, potash additions are more likely to be converted to slowly available reserves in the soil and it will be more difficult to raise the soil K index. However soil K targets and replacement principles for application remain the same as for other soils.

Other Factors Affecting Interpretation

Soil structure is also very important because any restriction to root growth may restrict the plants ability to obtain an adequate nutrient supply despite a satisfactory value indicated by analysis. The remedy is not to apply more nutrient but to improve the soil structure.

Organic matter-The level of organic matter (humus) will also affect the availability of nutrients in a soil and regular addition of manures so that the physical conditions and biological activity is improved will increase the plant-available nutrients.

Stone content can also have a large effect upon nutrient supply. Very stony soils have little fine earth yet it is the nutrients in this fine earth fraction that are measured. In consequence it is advisable to maintain very stony soils at slightly higher levels of available P and K than are required on deep stone-free soils. Soil depth-Crops frequently access some nutrients from below sampling depth and from subsoil (especially potash). Soils with greater rooting depth potential, and therefore volume, provide larger total quantities of nutrient than shallow soils. It may be possible to manage deep, well textured soils at slightly lower concentrations than the stated targets given above. However shallow rooting crops which do not explore the full soil volume will still require normal nutrient concentrations.

Changes in Soil Nutrient Status

Changes in nutrient status relate to the balance between nutrients applied and removed or lost (little P, K or Mg is leached from most properly managed UK soils). Nutrient balances per hectare are normally small especially in comparison with the total quantity of soil nutrients per hectare. Where the balance is positive not all the residual phosphate or potash applied remains as available soil P and K and therefore only small changes in soil status should be expected. Large changes, where not related to very large nutrient balances, need to be investigated.

Improving Soils with Low Nutrient Status

Soil analysis measures the average nutrient concentration in the depth of soil sampled and theoretically a concentration of 1 mg/l to a depth of 10 cm in soil represents 1 kg/ha of (elemental) nutrient. In practice however, plants also obtain nutrients from below sampling depth and from the slowly available pool that is not measured by analysis. This often gives rise to some confusion especially as these sources are very variable depending on soil type and past fertilisation and manuring. It is therefore possible to have a soil at 125 mg/l of K which supplies a cereal crop with a peak uptake of 250 kg/ha K (300 kg/ha K_2O).

Conversely the addition of 100 kg/ha of phosphorus, potassium or magnesium (in fertiliser terms this represents 230 kg P_2O_5 120 kg K_2O

and 170 kg MgO) will not result in an increase of 100 mg/l of readily plant-available soil P, K or Mg. This is because some of the nutrient applied will not remain in the readily available pool measured by analysis. Unfortunately the proportion remaining available and thus affecting soil analysis values will differ widely with different soils and conditions.

With present knowledge it is not possible to provide more than a very rough guide as to how soil values will change with additions (or removals) of nutrient. The following figures have been suggested but it is considered that the range in practice is even wider than indicated:

To increase soil Olsen P by 10mg/l requires 400-600 kg/ha P_2O_5

To increase plant-available soil K by 50mg/l requires 300-500kg/ha K_2O

More potash will be required on heavier soils where the clay type can make a difference. Where either the P or K status needs to be increased, triple superphosphate and muriate of potash (KCI) respectively are cost-effective nutrient sources.

Treatment of High Fertility Soils

No additions of nutrient (fertilisers or manures) should be made to any soils above index 4.

Soils at index 3 should receive maintenance dressings of phosphate but no potash for most arable crops (except potatoes) or grassland.

Some clay soils contain very large reserves of potash which are not reflected in the level of available K shown by normal analysis. In these cases the amount of slowly available K released each season is sufficient to replace some or all of the potash removed in combinable crop rotations, especially where the straw in not removed (the higher demand of roots and cut grass normally requires some addition). The soil K value may not change over many years cropping even though no potash is used to replace that removed. Not all heavy soils have this ability and where less nutrient is applied than is removed, soil levels must be regularly monitored for any reduction in plant-available K.

Conclusion

Whilst soil analysis is not a perfect tool, it is the most effective and practical means of assessing soil fertility in respect of pH and plant-available P, K and Mg. Analytical data should be used in conjunction with other knowledge such as soil type, structure and crop offtake as the basis for deciding on fertiliser and manure use.

Analysis and Collection of Soil Samples

Back in the nineteenth century Edmund Locard developed the theory known as the "Locard's Exchange Principle." The theory in short states that when ever someone comes in contact with another object or person there is a minute exchange of particles that, in theory, can be traced back to a victim or suspect (Block 1999). In a crime context, that evidence can confirm or disprove a hypothesis of involvement between a suspect and a victim. Locard described these particles as dust or dirt but today it is understood to include all soil borne trace evidence. Trace evidence can include blood, hair, fibre, dirt, glass particles and any other minute particles at a crime scene.

It would be nice to think that this forward thinker, Locard, moved beyond his Principle to developed the idea of soil analysis but he didn't. His mentor and friend Hans Gross published articles almost simultaneously with Sir Arthur Conan Doyle, the author of the Sherlock Holmes Fictions, regarding soil analysis (Block, 1999)(Nickel & Fischer, 1999) To Edmund Locard's great disappointment both Arthur Conan Doyle and Hans Gross beat him to the punch on the evaluation of soil as a forensic tool. However, Locard's zeal for the use of trace evidence in forensic investigations led to a life long study of the classification and identification of soil samples (Block, 1999). Since its dawn in the late 1800's the analysis of soil has grown into a multidisciplinary field of forensic study. Modern analysis of soil may involve geologists, entomologists, toxicologists, biologists, botanists and a myriad of other experts.

Soil may be understood as just that: soil. As such it has a value in forensic studies. However, soil may also be understood as a repository for non soil contaminants which can yield valuable information about crimes. In forensic analysis materials can be grouped in several ways and each lab has its own way of subdividing these groups (Chayko & Gulliver 1999). In general soil borne materials can be considered organic or inorganic. Both types are found in soil. Further refining of these classifications where soil is concerned is to break the groups into mineral, biological or synthetic matter.

Soil is considered trace evidence. Soil is made up of disintegrated surface material which can be organic, mineral or synthetic. The ratio of the mineral content compared to other matter in the soil can be very site specific. The ratios of mineral, organic and synthetic matter can vary even with in a few feet (Chayko & Gulliver, 1999). Sandy soils look, feel and behave quite differently from clay soils or peaty soils. By profiling an array of characteristics of each soil, it is somtimes possible to attribute those characteristics to a specific location (Steck, 2004).

Soil samples when properly taken can tell an investigator a lot about where a victim or suspect has been. Analysis of soil samples taken from vehicles can also tell an investigator about where a vehicle has been. Analysis of foot wear, clothing and tires can also place a suspect or victim in a particular location.

Collection of Samples

Collection of soil samples will depend on the circumstances of the crime. Indoor scenes will differ markedly from outdoor scenes in the type of evidence that can be recovered and the way in which these samples are collected.At indoor scenes there may be footprints in soil or in dust. Samples made by footwear should be photographed to scale before being recovered. The particle samples can best be collected using a vacuum method. The samples can be vacuumed with a portable vacuum cleaner equipped with a special attachment. The attachment has a metal screen on which a filter paper is attached. The area is vacuumed and the filter is removed and labeled with the date, location, time and name of the technician who operated the vacuum. The vacuum must be thoroughly cleaned between samples. Cleaning can be fairly easily done with handhelds where the parts are easy to access. Reference samples from the surrounding area perhaps including flower gardens, points of entry and exit and alibi locations should also be taken. Information on obtaining these specialized vacuum attachments can be obtained through Sirchie Laboratories Inc. in Youngsville N.C.

In the case of a break and enter or other crime at a home or business it is useful to know how the perpetrator entered the building. For example, if the perpetrator stood in a flower garden outside a window this indicates a stranger. If the perpetrator walked up the front lawn or to the back door this could indicate someone familiar with the property. Possibly the perpetrator knew no one was there being familiar with the owners habits. The soil on the shoes of a suspect (or left on a carpet or floor) can indicate the direction of travel and the mode of entry. This type of evidence is most useful when a suspect can be immediately identified before the soil is lost from the footwear.

Soil evidence from a victim or suspects clothing can indicate an association between victim and suspect. For instance, if a suspect lives in a particular neighborhood with a specific soil profile and this profile matches one found at a crime scene or on a victim, then one must suspect that the suspect left that sample at the scene.

"Double transfer" is even more convincing. If soil profiling to a suspects home territory is found on a victim or in their home and soil from the victims home territory is found on the suspects clothing or footwear the probability of the suspect having been at the crime scene increases dramatically. The mathematic probability of two matching profiles being found at both scenes by coincidence increases. For example the probability of transfer to one site occurring by coincidence might be 1 chance in 800. When double transfer has occurred the probability of that occurring by coincidence becomes 1 chance in 64,000 (Crocker, 1999).

Inorganic or manufactured matter found in the soil recovered from a suspect's footwear or clothing can be very site specific. Particles of glass, rubber and other industrial products can be used to link a suspect with a particular location. In the not-uncommon case of crimes at industrial locations this can be particularly useful.

How to Collect Samples

As already mention in the indoor scene vacuuming can be used to collect samples. The methods of collection will vary from scene to scene. In a break and enter where house plants are found upset a sample of the soil may be used to link the suspect to a scene. Soil from the garden or yard can also be used to track the suspect's direction of travel and point of entry. Most garden soils are unique. The gardener in charge will have added some favorite materials (sometimes specific to particular plants) in order to improve the garden. The materials tend to be different in every garden. Some gardeners might use compost while others might prefer ammonium pellets, sand, peat or wood chips.

Samples taken from the interior of a vehicle can indicate many things. A soil sample from the gas or brake pedal of a vehicle can link a suspect to a location. Soil from a trunk or backseat can indicate digging. When graves are dug to hide bodies the various levels of the soil are disturbed. If the suspect lays a shovel on the back seat or in the trunk the soil from the lower levels of the grave may be deposited on the seat or in the truck. Soil from tools such as shovels should be preserved in the state it was found. The entire tool should be packaged in protective material then enclosed in plastic. Finally the entire tool should be placed in a wooden or cardboard box and transported to the lab.

Sampling methods may need to be adapted when the scene is outdoors. If the challenge is the recovery of remains, soil samples should be taken at regular intervals up to 100 yards from the gravesite or point of recovery (Saferstein, 2004). Usually a grid search is set up and samples can be taken from each square of the grid and labeled as to which grid it was taken from.

About a tablespoon of soil should be enough for most modern tests. Usually only the surface soil needs to be sampled. The exception to this is in the case of a buried body, In this case soil samples should be taken at regular intervals as the remains are exposed. After the remains are removed from the gravesite the bottom of the grave should also be sampled. It is important to note that a new shovel, spoon or other scoop must be used for each grid and sample. Should the same implement is used serially, uncleaned to recover sample then Locard's Principle is at work again and cross contamination will render your samples useless.

Samples should be placed in plastic vials for transport. If it is not possible to transport the samples immediately they should be allowed to air dry before transport.

A special note for gravesites is that if insect evidence is present the vials should be labeled in pencil rather than pen. The specimens will likely be preserved in alcohol which if it leaks onto the label will destroy ink from pens. Without the label as to the time and place of collection your sample is forensically worthless.

Outdoor scenes often involve vehicles. There are several ways of collecting soil samples from vehicles. Vehicles involved in accidents will sometimes leave lumps of soil from under the wheel wells and fenders on the road way. These lumps should be collected intact and wrapped in protective material to minimize bumping during transport. The purpose behind this is to preserve the layers that have built up to create this lump of soil. A soil analyst can read these layers and know where this vehicle has been. The analyst may also be able to match the layers in a lump of soil to a particular vehicle. I can't help but think this would have been very useful in tracking Ted Bundy or Henry Lee Lucas in their cross county travels.

Clothing and footwear should be collected intact. No attempt should be made to remove soil from clothing, footwear or tires. If these items can be removed intact they should be placed in a paper bag or enclosed in a druggist's fold and then placed in a paper bag. Care must be taken so that the paper bag is protected so that evidence is not lost through holes in the bag. Some evidence can be placed in plastic bags but is must be completely dry. Wet samples placed in plastic with quickly degrade and rot and become useless. Plastic does not "breath" to allow passage of moisture and air. Paper on the other hand will allow moisture to escape preventing rot.

Lumps of soil stained with blood, semen or other biological samples should be collected intact and transported to the lab as dry samples.

Any samples containing suspected biological material such as blood, flesh, semen or hair must be clearly labeled so that the analyst at the lab can take precautions to preserve this material. Biological specimens in soil must never be heat dried.

The only time a sample should not be allowed to dry is when insect evidence such as maggots are present. In the case of maggots there is a very specific way of handling this evidence. Samples containing maggots can be placed in aluminum foil with a small piece beef liver and placed in a plastic container. There must be air available in the container. These samples must be sent to the lab immediately. The specimens can be placed in a thermally protected case such as a cooler. Attempts should be made to protect the samples from extremes of heat and cold. Never use dry ice with biological or entomological (insect) evidence.

In the Laboratory

In the lab items from the victim and suspect should be examined separately. Ideally the items and samples from the victim and suspect should be processed in different rooms. The personnel handling these sample should be assigned to one or the other. If this is not possible then the person handling the material should take extreme care to avoid cross contamination of the samples. The laboratory staff will evaluate the soil samples in several ways. First the mineral content will be tested. Some experienced analysts can moisten a sample and feel the soil. Based on feel alone they can tell the ratio of mineral and organic content. Microscopic examination of soil samples will subsequently reveal the type and nature of the mineral, biological and synthetic content of a sample.

Such talented analysts are not always available. Then, there are several standard testes which can identify the type and origin of the sample. The first and probably the most common is the density gradient tube. Two different liquids are added to a glass tube in various ratios. Each ratio represents a different density. The soil sample is poured into the tube. When the various particles reach a level in the liquid where their density is equal to the liquid the particles become suspended. This creates a unique profile of bands in the tube which can be matched to other samples.

The samples may also be tested using heat to test the point at which the sample will undergo an exothermic reaction or an endothermic reaction. The sample is heated in a special furnace to various temperatures. In an exothermic reaction the sample essentially burns and releases heat. In an endothermic reaction the sample will absorb the heat. Each sample from different locations will have these reactions at different temperatures according to the mineral, biological and synthetic content. Electron microscopes can be used to reveal the crystalline structures of minerals and synthetic material in a sample of soil. Nuclear Resonencing and Mass Spectometry are also methods which may be used in the Laboratory.

It is important that the crime investigators and the laboratory analyst communicate properly. Perhaps this happens best when the laboratory head is knowledgeable about investigations and is the contact person for investigators. The head can be briefed on the questions and issues of the case and then direct the laboratory personnel as to the direction of the inquiry. The investigators may want to know if sample #12 and sample #76 are similar. The laboratory head will then choose appropriate methods which express the similarity between samples. Once the questions and procedures are chosen everything depends on the integrity of the sample. If they have been collected with care and documented amply then the results from the laboratory can be trusted (Steck, 2004).

Conclusion

Take samples of all the soil in and around a crime scene. Place a reference sample in a plastic vial and label it with the date the time the investigators name and the case number. Reference samples are samples of soil from places the suspect or victim may have picked up soil. The reference sample can also be from sites the suspect may have been. Include the location and the distance from the focal point of the crime when labeling samples.

Reference samples should be about a tablespoon or so taken from less than a 1/2 an inch (about 1 cm). The exception is in cases of burial where samples should be taken every 1 inch (every 3cm). With gravesites the bottom of the grave should be sampled after the removal of the remains. Package and label all samples carefully. Keep in mind that the analysts in the laboratory do not have firsthand knowledge of the scene and the circumstances of collection. Try to provide a detailed description of the scene and the circumstances of the crime which the laboratory staff can use to determine what type of analysis is most appropriate.

Water Quality Assessment

Concerns and Limitations

Salts exert both general and specific effects on plants which directly influence crop yield. Additionally, salts affect certain soil physico-chemical properties which, in turn, may affect the suitability of the soil as a medium for plant growth. The development of appropriate criteria and standards for judging the suitability of a saline water for irrigation and for selecting appropriate salinity control practices requires relevant knowledge of how salts affect soils and plants. This section presents a brief summary of the principal salinity effects that should be thoroughly understood in this regard.

Effects of Salts on Soils

The suitability of soils for cropping depends heavily on the readiness with which they conduct water and air (permeability) and on aggregate properties which control the friability of the seedbed (tilth). Poor permeability and tilth are often major problems in irrigated lands. Contrary to saline soils, sodic soils may have greatly reduced permeability and poorer tilth. This comes about because of certain physico-chemical reactions associated, in large part, with the colloidal fraction of soils which are primarily manifested in the slaking of aggregates and in the swelling and dispersion of clay minerals.

To understand how the poor physical properties of sodic soils are developed, one must look to the binding mechanisms involving the negatively charged colloidal clays and organic matter of the soil and the associated envelope of electrostatically adsorbed cations around the colloids, and to the means by which exchangeable sodium, electrolyte concentration and pH affect this association. The cations in the "envelope" are subject to two opposing processes:

- they are attracted to the negatively-charged clay and organic matter surfaces by electrostatic forces, and
- they tend to diffuse away from these surfaces, where their concentration is higher, into the bulk of the solution, where their concentration is generally lower.

The two opposing processes result in an approximately exponential decrease in cation concentration with distance from the clay surfaces into the bulk solution. Divalent cations, like calcium and magnesium, are attracted by the negatively-charged surfaces with a force twice as great as monovalent cations like sodium. Thus, the cation envelope in the divalent system is more compressed toward the particle surfaces. The envelope is also compressed by an increase in the electrolyte concentration of the bulk solution, since the tendency of the cations to diffuse away from the surfaces is reduced as the concentration gradient is reduced.

The associations of individual clay particles and organic matter micelles with themselves, each other and with other soil particles to form assemblages called aggregates are diminished when the cation "envelope" is expanded (with reference to the surface of the particle) and are enhanced when it is compressed. The like-electrostatic charges of the particles which repel one another and the opposite-electrostatic charges which attract one another are relatively long-range in effect. On the other hand, the adhesive forces, called Vanderwaal forces, and chemical bonding reactions involved in the particle-to-particle associations which bind such units into assemblages, are relatively short-range forces. The greater the compression of the cation "envelope" toward the particle surface, the smaller the overlap of the "envelopes" and the repulsion between adjacent particles for a given distance between them. Consequently, the particles can approach one another closely enough to permit the adhesive forces to dominate and assemblages (aggregates) to form.

The phenomenon of repulsion between particles causes more soil solution to be imbibed between them (this is called swelling). Because clay particles are plate-like in shape and tend to be arranged in parallel orientation with respect to one another, swelling reduces the size of the inter-aggregate pore spaces in the soil and, hence, permeability. Swelling is primarily important in soils which contain substantial amounts of expanding-layer phyllosilicate clay minerals (smectites like montmorillonite) and which have ESP values in excess of about 15. The reason for this is that, in such minerals, the sodium ions in the pore fluid are first. attracted to the external surfaces of the clay plate. Only after satisfying this do the sodium ions occupy the space between the parallel platelets of the oriented and associated clay particles of the sub-aggregates (called domains) where they create the repulsion forces between adjacent platelets which lead to swelling.

Dispersion (release of individual clay platelets from aggregates) and slaking (breakdown of aggregates into subaggregate assemblages) can occur at relatively low ESP values (<15), provided the electrolyte concentration is sufficiently low. The packing of aggregates is more porous than that of individual particles or subaggregates, hence permeability and tilth are better in aggregated conditions. Repulsed clay platelets or slaked subaggregate assembles can lodge in pore interstices, also reducing permeability.

Thus, soil solutions composed of high solute concentrations (salinity), or dominated by calcium and magnesium salts, are conducive to good soil physical properties. Conversely, low salt concentrations and relatively high proportions of sodium salts adversely affect permeability and tilth. High pH (> 8) also adversely affects permeability and tilth because it enhances the negative charge of soil clay and organic matter and, hence, the repulsive forces between them.

During an infiltration event, the soil solution of the topsoil is essentially that of the infiltrating water and the exchangeable sodium percentage is essentially that pre-existent in the soil (since ESP is buffered against rapid change by the soil cation exchange capacity). Because all water entering the soil must pass through the soil surface, which is most subject to loss of aggregation, topsoil properties largely control the water entry rate of the soil.

These observations taken together with knowledge of the effects of the processes discussed above explain why soil permeability and tilth problems must be assessed in terms of both the salinity of the infiltrating water and the exchangeable sodium percentage (or its equivalent SAR value) and the pH of the topsoil. Representative threshold values of SAR (- ESP) and the electrical conductivity of infiltrating water for maintenance of soil permeability.

Because there are significant differences among soils in their susceptibilities in this regard, this relation should only be used as a guideline. The data available on the effect of pH are not yet extensive enough to develop the third axis relation needed to refine this guideline (Suarez *et al.* 1984; Goldberg and Forster 1990; Goldberg *et al.* 1990).

Decreases in the infiltration rate (IR) of a soil generally occur over the irrigation season because of the gradual deterioration of the soil's structure and the formation of a surface seal (horizontally layered arrangement of discrete soil particles) created during successive irrigations (sedimentation, wetting and drying events). IR is even more sensitive to exchangeable sodium, electrolyte concentration and pH than is hydraulic conductivity.

This is due to the increased vulnerability of the topsoil to mechanical forces, which enhance clay dispersion, aggregate slaking and the movement of clay in the "loose" near-surface soil, and to the lower electrolyte concentration that generally exists there, especially under conditions of rainfall.

Depositional crusts often form in the furrows of irrigated soils where soil particles suspended in water are deposited as the water flow rate slows or the water infiltrates. The hydraulic conductivity of such crusts is often two to three orders of magnitude lower than that of the underlying bulk soil, especially when the electrolyte concentration of the infiltrating water is low and exchangeable sodium is relatively high.

The addition of gypsum (either to the soil or water) can often help appreciably in avoiding or alleviating problems of reduced infiltration rate and hydraulic conductivity.

For more specific information on the effects of exchangeable sodium, electrolyte concentration and pH, as well as of exchangeable Mg and K, and use of amendments on the permeability and infiltration rate of soils reference should be made to the reviews of Keren and Shainberg (1984); Shainberg (1984); Emerson (1984); Shainberg and Letey (1984); Shainberg and Singer (1990).

Effects of Salts on Plants

Excess salinity within the plant rootzone has a general deleterious effect on plant growth which is manifested as nearly equivalent reductions in the transpiration and growth rates (including cell enlargement and the synthesis of metabolites and structural compounds).

This effect is primarily related to total electrolyte concentration and is largely independent of specific solute composition. The hypothesis that best seems to fit observations is that excessive salinity reduces plant growth primarily because it increases the energy that must be expended to acquire water from the soil of the rootzone and to make the biochemical adjustments necessary to survive under stress.

This energy is diverted from the processes which lead to growth and yield. Growth suppression is typically initiated at some threshold value of salinity, which varies with crop tolerance and external environmental factors which influence the need of the plant for water, especially the evaporative demand of the atmosphere (temperature, relative humidity, windspeed, etc.) and the water-supplying potential of the rootzone, and increases as salinity increases until the plant dies. The salt tolerances of various crops are conventionally expressed (after Maas and Hoffman 1977), in terms of relative yield (Y_r), threshold salinity value (a), and percentage decrement value per unit increase of salinity in excess of the threshold (b); where soil salinity is expressed in terms of EC_e, in dS/m), as follows:

$$Y_r = 100\text{-}b\ (EC_e\text{-}a)$$

Where Y_r- is the percentage of the yield of the crop grown under saline conditions relative to that obtained under non-saline, but otherwise comparable, conditions. This use of EC_e to express the effect of salinity on yield implies that crops respond primarily to the osmotic potential of the soil solution. Tolerances to specific ions or elements are considered separately, where appropriate.

Table: *Relative salt tolerance of various crops at emergence and during growth to maturity (after Maas 1986)*

Crop Electrical conductivity of saturated soil extract

Common name	*Botanical name*[1]	*50% yield dS/m*	*50% emergence*[2] *dS/m*
Barley	*Hordeum vulgare*	18	16-24
Cotton	*Gossypium hirsutum*	17	15
Sugarbeet	*Beta vulgaris*	15	6-12
Sorghum	*Sorghum bicolor*	15	13
Safflower	*Carthamus tinctorius*	14	12
Wheat	*Triticum aestivum*	13	14-16
Beet, red	*Beta vulgaris*	9.6	13.8
Cowpea	*Vigna unguiculata*	9.1	16
Alfalfa	*Medicago sativa*	8.9	8-13
Tomato	*Lycopersicon lycopersicum*	7.6	7.6
Cabbage	*Brassica oleracea capitata*	7.0	13
Maize	*Zea mays*	5.9	21-24
Lettuce	*Lactuca sativa*	5.2	11
Onion	*A/Hum cepa*	4.3	5.6-7.5
Rice	*Oryza sativa*	3.6	18
Bean	*Phaseolus vulgaris*	3.6	8.0

1 Botanical and common names follow the convention of Hortus Third where possible.

2 Emergence percentage of saline treatments determined when non-saline treatments attained maximum emergence.

It is important to recognize that such salt tolerance data cannot provide accurate, quantitative crop yield losses from salinity for every situation, since actual response to salinity varies with other conditions of growth including climatic and soil conditions, agronomic and irrigation management, crop variety, stage of growth, etc.

While the values are not exact, since they incorporate interactions between salinity and the other factors, they can be used to predict how one crop might fare relative to another under saline conditions.

Climate is a major factor affecting salt tolerance; most crops can tolerate greater salt stress if the weather is cool and humid than if it is hot and dry. Yield is reduced more by salinity when atmospheric humidity is low. Ozone decreases the yield of crops more under non-saline than saline conditions, thus the effects of ozone and humidity increase the apparent salt tolerance of certain crops.

Plants are generally relatively tolerant during germination but become more sensitive during emergence and early seedling stages of growth; hence it is imperative to keep salinity in the seedbed low at these times.

Significant differences in salt tolerance occur among varieties of some species though this issue is confused because of the different climatic or nutritional conditions under which the crops were tested and the possibility of better varietal adaption in this regard. Rootstocks affect the salt tolerances of tree and vine crops because they affect the ability of the plant to extract soil water and the uptake and translocation to the shoots of the potentially toxic sodium and chloride salts.

Salt tolerance also depends somewhat upon the type, method and frequency of irrigation. As the soil dries, plants experience matric stresses, as well as osmotic stresses, which also limit water uptake. The prevalent salt tolerance data apply most directly to crops irrigated by surface (furrow and flood) methods and conventional irrigation management. Salt concentrations may differ several-fold within irrigated soil profiles and they change constantly. The plant is most responsive to salinity in that part of the rootzone where most of the water uptake occurs. Therefore, ideally, tolerance should be related to salinity weighted over time and measured where the roots absorb most of the water.

Sprinkler-irrigated crops are potentially subject to additional damage caused by foliar salt uptake and desiccation (burn) from spray contact of the foliage. For example, Bernstein and Francois (1973a) found that the yields of bell peppers were reduced by 59 percent more when 4.4 dS/m water was applied by sprinklers compared to a drip system. Meiri (1984) found similar results for potatoes.

Susceptibility of plants to foliar salt injury depends on leaf characteristics affecting rate of absorption and is not generally correlated with tolerance to soil salinity. The degree of spray injury varies with weather conditions, especially the water deficit of the atmosphere. Visible symptoms may appear suddenly following irrigations when the weather is hot and dry. Increased frequency of sprinkling, in addition to increased temperature and evaporation, leads to increases in salt concentration in the leaves and in foliar damage.

While the primary effect of soil salinity on herbaceous crops is one of retarding growth, as discussed above, certain salt constituents are specifically toxic to some crops. Boron is such a solute and, when present in the soil solution at concentrations of only a few mg/l, is highly toxic to susceptible crops. Boron toxicities may also be described in terms of a threshold value and yield-decrement slope parameters, as is salinity.

Generally these plants are also salt-sensitive and the two effects are difficult to separate.

Sodic soil conditions may induce calcium, as well as other nutrient, deficiencies because the associated high pH and bicarbonate conditions repress the solubilities of many soil minerals, hence limiting nutrient concentrations in solution and, thus, availability to the plant.

Table: *Relative susceptibility of crops to foliar injury from saline sprinkling water*[1] *(after Maas 1990)*

Na or Cl conc ($mmol_c/l$) causing foliar injury[2]			
<5	***5-10***	***10-20***	***>20***
Almond	Grape	Alfalfa	Cauliflower
Apricot	Pepper	Barley	Cotton
Citrus	Potato	Cucumber	Sugarbeet
Plum	Tomato	Maize	Sunflower
		Safflower	
		Sesame	
		Sorghum	

[1] Susceptibility based on direct accumulation of salts through the leaves.

[2] Foliar injury is influenced by cultural and environmental conditions. These data are presented only as general guidelines for day-time sprinkling.

Table: *Boron tolerance limits for agricultural crops (after Maas 1990)*

Common name	***Botanical name***	***Threshold[1] g/m^3***	***Slope % per g/m^3***
Very sensitive			
Lemon[2]	*Citrus limon*	<0.5	
Blackberry[2]	*Rubus* sp.	<0.5	
Sensitive			
Avocado[2]	*Persea americana*	0.5-7.5	
Grapefruit[2]	*C.* × *paradisi*	0.5-7.5	
Orange[2]	*C. sinensis*	0.5-7.5	
Apricot[2]	*Prunus armeniaca*	0.5-7.5	
Peach[2]	*P. persica*	0.5-7.5	
Cherry[2]	*P. avium*	0.5-7.5	
Plum[2]	*P. domestica*	0.5-7.5	
Persimmon[2]	*Diospyros kaki*	0.5-7.5	
Fig, kadota[2]	*Ficus carica*	0.5-7.5	
Grape[2]	*Vitis vinifera*	0.5-7.5	
Walnut[2]	*Juglans regia*	0.5-7.5	

Contd...

Common name	Botanical name	Threshold[1] g/m^3	Slope % per g/m^3
Pecan[2]	*Carya illinoiensis*	0.5-7.5	
Onion	*Allium cepa*	0.5-7.5	
Garlic	*A. sativum*	0.75-1.0	
Sweet potato	*Ipomoea batatas*	0.75-1.0	
Wheat	*Triticum aestivum*	0.75-1.0	3.3
Sunflower	*Helianthus annuus*	0.75-1.0	
Bean, mung[2]	*Vigna radiata*	0.75-1.0	
Sesame[2]	*Sesamum indicum*	0.75-1.0	
Lupine[2]	*Lupinus hartwegii*	0.75-1.0	
Strawberry[2]	*Fragaria sp.*	0.75-1.0	
Artichoke, Jerusalem[2]	*Helianthus tuberosus*	0.75-1.0	
Bean, kidney[2]	*Phaseolus vulgaris*	0.75-1.0	
Bean, snap	*P. vulgaris*	1.0	12
Bean, lima[2]	*P. lunatus*	0.75-1.0	
Groundnut	*Arachis hypogaea*	0.75-1.0	
Moderately tolerant			
Broccoli	*Brassica oleracea botrytis*	1.0	1.8
Pepper, red	*Capsicum annuum*	1.0-2.0	
Pea[2]	*Pisum sativa*	1.0-2.0	
Carrot	*Daucus carota*	1.0-2.0	
Radish	*Raphanus sativus*	1.0	1.4
Potato	*Solarium tuberosum*	1.0-2.0	
Cucumber	*Cucumis sativus*	1.0-2.0	
Lettuce	*Lactuca sativa*	1.3	1.7
Cabbage[2]	*Brassica oleracea capitata*	2.0-4.0	
Turnip	*B. rapa*	2.0-4.0	
Bluegrass, Kentucky[2]	*Poa pratensis*	2.0-4.0	
Barley	*Hordeum vulgare*	3.4	4.4
Cowpea	*Vigna unguiculata*	2.5	12
Oats	*Avena sativa*	2.0-4.0	
Maize	*Zea mays*	2.0-4.0	
Artichoke[2]	*Cynara scolymus*	2.0-4.0	
Tobacco[2]	*Nicotiana tabacum*	2.0-4.0	
Mustard[2]	*Brassica juncea*	2.0-4.0	
Clover, sweet[2]	*Melilotus indica*	2.0-4.0	

Contd...

Common name	Botanical name	Threshold[1] g/m^3	Slope % per g/m^3
Squash	*Cucurbita pepo*	2.0-4.0	
Muskmelon[2]	*Cucumis melo*	2.0-4.0	
Cauliflower	*B. olearacea botrytis*	4.0	1.9
Tolerant			
Alfalfa[2]	*Medicago sativa*	4.0-6.0	
Vetch, purple[2]	*Vicia benghalensis*	4.0-6.0	
Parsley[2]	*Petroselinum crispum*	4.0-6.0	
Beet, red	*Beta vulgaris*	4.0-6.0	
Sugarbeet	*B. vulgaris*	4.9	4.1
Tomato	*Lycopersicon lycopersicum*	5.7	3.4
Very tolerant			
Sorghum	*Sorghum bicolor*	7.4	4.7
Cotton	*Gossypium hirsutum*	6.0-10.0	
Celery[2]	*Apium graveolens*	9.8	3.2
Asparagus[2]	*Asparagus officinalis*	10.0-15.0	

[1] Maximum permissible concentration in soil water without yield reduction. Boron tolerances may vary, depending upon climate, soil conditions and crop varieties.

[2] Tolerance based on reductions in vegetative growth.

These conditions can be improved through the use of certain amendments such as gypsum and sulphuric acid. Sodic soils are of less extent than saline soils in most irrigated lands. Crops grown on fertile soil may seem more salt tolerant than those grown with adequate fertility, because fertility is the primary factor limiting growth. However, the addition of extra fertilizer will not alleviate growth inhibition by salinity. For a more thorough treatise on the effects of salinity on the physiology and biochemistry of plants, see the reviews of Maas and Nieman (1978), Maas (1990) and Lauchli and Epstein (1990).

Table: *Boron tolerances for ornamentals[1] (after Maas 1990)*

Common name	Botanical name	Threshold[2] mg/l
Very sensitive		
Oregon grape	*Mahonia aquifolium*	<0.5
Photinia	*Photinia × fraseri*	<0.5
Xylosma	*Xylosma congestum*	<0.5
Thorny elaeagnus	*Elaeagnus pungens*	<0.5

Contd...

Common name	***Botanical name***	***Threshold[2] mg/l***
Laurustinus	*Viburnum tinus*	<0.5
Wax-leaf privet	*Ligustrum japonicum*	<0.5
Pineapple guava	*Feijoa sellowiana*	<0.5
Spindle tree	*Euonymus japonica*	<0.5
Japanese pittosporum	*Pittosporum tobira*	<0.5
Chinese holly	*Ilex cornuta*	<0.5
Juniper	*Juniperus chinensis*	<0.5
Yellow sage	*Lantana camara*	<0.5
American elm	*Ulmus americana*	<0.5
Sensitive		
Zinnia	*Zinnia eleganus*	0.5-1.0
Pansy	*Viola tricolor*	0.5-1.0
Violet	*V. odorata*	0.5-1.0
Larkspur	*Delphinium* sp.	0.5-1.0
Glossy abelia	*Abelia × grandiflora*	0.5-1.0
Rosemary	*Rosmarinus officinalis*	0.5-1.0
Oriental arbovitae	*Platycladus orientalis*	0.5-1.0
Geranium	*Pelargonium × hortorum*	0.5-1.0
Moderately sensitive		
Gladiolus	*Gladiolus* sp.	1.0-2.0
Marigold	*Calendula officinalis*	1.0-2.0
Poinsettia	*Euphorbia pulcherrima*	1.0-2.0
China aster	*Callistephus chinensis*	1.0-2.0
Gardenia	*Gardenia* sp.	1.0-2.0
Southern yew	*Podocarpus marcophyllus*	1.0-2.0
Brush cherry	*Syzygium paniculatum*	1.0-2.0
Blue dracaena	*Cordyline indivisa*	1.0-2.0
Ceniza	*Leucophyllus frutescens*	1.0-2.0
Moderately tolerant		
Bottlebrush	*Callistemon citrinus*	2.0-4.0
California poppy	*Eschscholzia californica*	2.0-4.0
Japanese boxwood	*Buxus microphylla*	2.0-4.0
Oleander	*Nerium oleander*	2.0-4.0
Chinese hibiscus	*Hibiscus rosa-senensis*	2.0-4.0
Sweet pea	*Lathyrus odoratus*	2.0-4.0
Carnation	*Dianthus caryophyllus*	2.0-4.0

Contd...

Common name	*Botanical name*	*Threshold[2] mg/l*
Tolerant		
Indian hawthorn	*Raphiolephis indica*	6.0-8.0
Natal palm	*Carissa grandiflora*	6.0-8.0
Oxalis	*Oxalis bowiei*	6.0-8.0

[1] Species listed in order of increasing tolerance based on appearance as well as growth reduction.

[2] Boron concentrations exceeding the threshold may cause leaf burn and loss of leaves.

Effects of Salts on Crop Quality

Information on the effects of water salinity and/or soil salinity on crop quality is very scant although such effects are apparent and have been noticed under field conditions. In general, soil salinity, either caused by saline irrigation water or by a combination of water, soil and crop management factors, may result in: reduction in size of the produce; change in colour and appearance; and change in the composition of the produce. Shalhevet *et al.* (1969) reported a reduction of seed size in groundnuts beginning at soil salinity levels (EC_e) of 3 dS/m. However, there is an increase in seed oil content with increasing salinity up to a point. In the case of tomatoes, it was reported (Shalhevet and Yaron 1973) that for every increase in 1.5 dS/m in mean EC_e beyond 2 dS/m, there was a 10 percent reduction in yield. The yield reduction was due only to reduction in fruit size and weight and not to reduction in fruit number. However, there was a marked increase in soluble solids in the extract, which may be an important criterion for tomato juice production. If ever tomato juice processors purchase tomatoes on the basis of total solids content, there would be no economic penalty for salinity in the range up to 6.0 dS/m in EC_e.

The mean pH of the juice was 4.3 with no meaningful differences among treatments. Fruits from higher salinity treatments were less liable to damage and the number of spoiled fruits was less.

Table: *Effect of soil salinity on seed weight and oil content in groundnuts (Shalhevet* et at. *1969)*

EC_e dS/m	*Weight of 1000 seeds, g*	*Oil content % dry weight*
1.74	774	48.9
2.92	690	49.0
3.16	676	50.2
4.41	656	47.6
5.61	470	46.2

Table: *Effect of soil salinity on fruit weight and soluble solid content of tomatoes*

EC_e *dS/m*	*Weight per fruit g*	*% soluble solids*	*% spoiled fruits*
1.6	68.5	4.5	15.5
3.8	59.5	4.5	17.7
6.0	55.8	4.8	12.3
10.2	51.9.	5.9	11.1

Meiri *et al.* (1981) reported that increased salinity reduced fruit size in muskmelons (*Cucumis melo*). However, ripening was accelerated by salinity.

Bielorai *et al.* (1978) reported that grapefruit yield decreased with increase in chloride ion concentration; the yield reduction was caused more by reduction in fruit size and weight. Salinity effects on fruit quality were similar to those caused by water stress. Comparing the low and high salinity levels, there is an increase in soluble solids and tritratable acidity in the juice. There were no differences in juice content. Rhoades *et al.* (1989) obtained increases in the quality of wheat, melons and alfalfa from use of saline drainage water for irrigation.

Water Quality Improvement: Institutional and Financial Solutions

Water quality problems and issues in southwestern Pennsylvania are both local and regional as evidenced by a, water quality assessments by the Pennsylvania Department of Environmental Protection (PADEP), and testimony received by the committee. Some of these water quality problems are associated primarily with urbanization in the immediate Pittsburgh vicinity; some are associated with activity in the Monongahela and Allegheny River basins; still others are common to the predominantly rural counties in southwestern Pennsylvania. Large differences exist among the sources of problems, their potential effects on public health and environmental quality, and their likely solutions. Further, resolution of water quality issues in southwestern Pennsylvania is affected by other regional issues such as transportation, land use, and governance of the metropolitan area.

The existing pattern of water supply and water quality services in the region is highly fragmented, with more than 1,000 providers operating in the multicounty region. In Pittsburgh's metropolitan area, like many other metro areas in the United States, large-special purpose authorities such as the Allegheny County Sanitary Authority (ALCOSAN) can achieve substantial economies of scale through joint management agencies. Although private organizations may not have direct voting power in what mix of organizations is chosen to implement the plan, they could very well influence how the public and its elected and appointed representatives make these choices.

It is desirable to have some mechanism to facilitate continued oversight of regional progress (or lack thereof) toward clean water and its relationships to other regional goals and activities, and to help southwestern Pennsylvania realize the benefits of cooperation.

Furthermore, the situation is not static. Although the Pittsburgh metropolitan statistical area (MSA) is among the few in the nation to actually lose population during the 1990s, it is nevertheless listed by American Rivers (2002) as among the top 20 metropolitan areas in terms of "urban sprawl."

This ranking is based on the percentage increase in developed land in 1997 compared to 1982. According to American Rivers, the Pittsburgh MSA experienced an increase of 42.5 percent in urbanized land, accompanied by a decrease in average density of 35.5 percent over those 15 years. Planning for water quality improvement, especially where capital investment is substantial, must therefore reflect regional planning goals concerning economic development and demographic character, such as impacts of urban sprawl and (re)development.

Finding the right mix of existing and new organizations that best fulfill the necessary conditions for planning, implementation, and oversight of CWARP will be a difficult and time-consuming process. Several options that the region should consider are discussed in this chapter. The discussion begins with a review of management functions necessary to deliver water supply and water quality services and criteria for evaluating alternative organizational arrangements to perform those functions. The challenge is to find the right mix of organizations that can perform the necessary functions in an efficient and politically accountable manner.

The committee's examination of specific arrangements begins with existing organizations in the region. This is followed by a brief review of what other regions with somewhat similar problems have done. Future options for water resource and quality management in southwestern Pennsylvania are then explored. These options are discussed in light of existing enabling legislation and what additional legislation may be desirable. Also, two other significant factors influencing the choice of organizational arrangements are discussed: (1) potential sources of financing and (2) financial burdens that may be imposed on citizens of the region.

Criteria for Evaluating Organizational Options

Choosing an appropriate organization or set of organizations to address regional water quality problems holistically is a complex task. Criteria for guiding the formulation and evaluation of alternative arrangements usually include consideration of the following:

- efficiencies with which each organizational arrangement could carry out the various policy-making and management functions by exploiting economies of scale;
- geographic coverage sufficient to incorporate significant hydrological, biological, and chemical processes between upstream and downstream elements of the water resource system and to incorporate significant linkages in construction and operation of infrastructure that crosses political boundaries;
- capacity to integrate water systems, wastewater systems, stormwater systems, and other aspects of water resources with land use and transportation;
- legal, technical, and financial capacities of each option to perform management functions;
- capacity of each option to involve the many faces of the public and minimize conflict in decision making processes; and
- the nature of existing contracts and other commitments.

Before these criteria can meaningfully be applied, it is appropriate to describe the management functions, scale, and authorities of alternative arrangements.

Management Functions

A list of water quality planning and management functions for water systems. They are listed in approximate order of statutory authority necessary to perform them, beginning with the least intrusive government power and concluding with the most intrusive. Collection of data, planning, and technical assistance require only modest statutory authority. Implementing actions including financing, construction, taking of land, and adoption and enforcement of regulations require substantially greater authority.

General-purpose local governments, including municipalities and counties, usually have the broadest array of powers delegated to them by state legislatures. Therefore, they tend to face fewer legal obstacles, exercise greater power to integrate land use and water services, and have greater flexibility to implement economically efficient management programs within their limited geographical jurisdictions.

Issues of Scale

Scale is a key factor in selecting an appropriate mix of organizations to deliver services in the region. The National Research Council (NRC) Committee on Watershed Management (NRC, 1999) addressed the issue of choosing an appropriate scale for planning that includes all relevant hydrologic linkages, commenting as follows:

Managing water resources at the watershed scale, while difficult, offers the potential of balancing the many, sometimes competing, demands we place on water resources. The watershed approach acknowledges linkages between upland and downstream areas, and between surface and ground water, and reduces the chances that attempts to solve problems in one realm will cause problems in other...Organizations for watershed management are most likely to be effective if their structure matches the scale of the problem.

Planning at the watershed scale offers the opportunity to address externalities among several parties within the basin. That earlier NRC committee addressed the problem of incorporating hydrologic and biological interdependencies that exist in water resource systems. Unfortunately, the geographic jurisdictions of organizations with the range of necessary legal authorities seldom match watershed boundaries.

There are approximately are parts of five counties and 100 municipalities within that area alone. New organizational arrangements may have to be created to effectively and efficiently manage water, but development of these arrangements may entail difficult political decisions that involve the transfer of some powers and responsibilities from existing units of government. These difficulties

must be weighed against the anticipated economy-of-scale benefits that new organization(s) may offer.

Planning and management are needed to address the array of water resource problems at four interrelated scales in (and beyond) the Pittsburgh region, and organizational arrangements should be responsive to each of the following scales:

1. river basin, to address issues related to imports and exports to the multicounty region, including areas and states outside southwestern Pennsylvania;
2. multicounty/metropolitan scale, where decisions are being made about large-scale infrastructure and related land use in southwestern Pennsylvania that affect water resources and where opportunities exist to achieve efficiencies and avoid conflicts in regional water management;

3. urban areas in and around Allegheny County and outlying urban centers, where combined and separate sewer overflows and stormwater runoff must be addressed; and
4. rural areas within southwestern Pennsylvania having problems of inadequate human waste disposal and water supply.

Current Situation in Southwestern Pennsylvania

Water quality management in the Pittsburgh region is highly fragmented, with responsibilities and authority distributed among a very large number of general purpose local governments, special districts, regional planning organizations, and the Commonwealth of Pennsylvania. For purposes of this discussion, the region is defined by the nine-county area served by the Southwestern Pennsylvania Commission (SPC). It is important to note that alternative definitions are discussed elsewhere in this report.

General Purpose Local Governments and Special Districts

The 2002 Census of Governments lists 526 general purpose governments within the region, distributed by county and type of government. In the 1997 Census of Governments, boroughs, cities, and municipalities were lumped together under the heading "cities" (the number of cities in the 1997 census is the same as the sum of boroughs plus municipalities plus cities in the 2002 census), and the numbers were unchanged from 1997 to 2002. Under Pennsylvania law, each of those local governments and the nine counties are authorized to provide water supply and sewer services.

Table: *General Purpose Local Governments in Southwestern Pennsylvania in 2002*

County	*Cities and Boroughs*	*Municipalities*	*Townships*
Allegheny	80	6	42
Armstrong	16	1	28
Beaver	27	2	22
Butler	23	1	33
Fayette	16	2	24
Greene	6	0	20
Indiana	14	0	24
Washington	33	2	32
Westmoreland	36	8	21
Total	251	22	246

Source: United States Census of Governments, 2002,

In addition to the general purpose governments, there are 154 special districts engaged in either sewer service alone or both water supply and sewer service. The special districts are distributed by county, type, and characteristics of service boundaries.

The only special districts included in the 1997 list of "large" districts in the Census of Government finances that were clearly identifiable as delivering sewer services were the Pittsburgh Water and Sewer Authority, with an annual expenditure of about $118 million, and ALCOSAN, with expenditures of $284 million annually. ALCOSAN serves 83 communities, most of which are located in or immediately adjacent to Allegheny County. Fragmentation of sewer services in the region with its many special districts reflects the general pattern of special districts in Pennsylvania. The 1997 Census of Governments reported 2,004 single-purpose sewer districts in the United States; Pennsylvania had the highest number, 591, about 30 percent of the nation's total. Wisconsin was the next highest state with 320 single-purpose sewer districts.

Regional Planning Organizations

The Southwestern Pennsylvania Commission (SPC) is the officially designated regional planning agency for the area in and around Pittsburgh. SPC's major role is "comprehensive regional planning with emphasis on transportation and economic development." It was designated in 1974 as the metropolitan planning organization for transportation. It is also the Economic Development District for southwestern Pennsylvania, as designated by the U.S. Appalachian Regional Commission and the U.S. Department of Commerce. The SPC governing board includes more than 60 members representing the 10 counties, the City of Pittsburgh, the Governor's Office, and several state and federal agencies.

In addition to its primary functions, recent discussions regarding regional land use and growth decisions have pointed to the need for SPC to help address local development issues (e.g., WSIP, 2002). As a result, SPC is expected to continue to create, organize, and support public forums that bring a regional perspective to issues such as housing, sewer systems, and community development.

In 1998, SPC requested that the Western Division of the Pennsylvania Economy League make a preliminary study of the region's needs. That study pointed to water supply and wastewater problems as potential impediments to future economic growth, and in 1999, the Western Division of the Pennsylvania Economy League initiated the Southwestern Pennsylvania Water and Sewer

Infrastructure Project (WSIP). The steering committee for that project included 60 public and private sector leaders from the region. As described elsewhere in this report, the WSIP report identifies several important water supply and wastewater management problems in the region, including the following:

- overflowing sewers and failing septic systems that annually discharge billions of gallons of inadequately treated or untreated sewage into the region's streams and lakes;
- lack of clean and reliable water supplies to some residents, particularly in rural areas;
- inadequate water and sewer infrastructure at otherwise desirable development sites; and
- growth limitations in many communities resulting from inadequate facilities.

The WSIP Steering Committee recommended the following:

- the SPC serve as the organization for setting regional water-related goals and priorities;
- that the Three Rivers Wet Weather Demonstration Program (3RWW) serve as the regional organization for public education and technical assistance, expanding its service area beyond the ALCOSAN area that it now serves; and
- that the Southwestern Pennsylvania Growth Alliance and the Greater Pittsburgh Chamber of Commerce serve as a regional advocacy organization.

Commonwealth of Pennsylvania

The Pennsylvania Department of Environmental Protection (PADEP) is the state regulatory agency charged with water quality management. In that capacity it has jurisdiction over those portions of the Ohio River basin within Pennsylvania, including the Allegheny and Monongahela River tributaries. The PADEP has included the Ohio River basin among six major basins in the state (the others being Lake Erie, Genessee, Susquehanna, Potomac, and Delaware). Unlike water resource planning under Pennsylvania's Water Resources Planning Act (WRPA) of 2002 (General Assembly of Pennsylvania, 2002), PADEP does not have a planning program to guide management of water quality at the basinwide scale.

The PADEP has, however, established a watershed restoration program at a smaller scale than the Ohio River basin under its nonpoint source program—Pennsylvania's response to requirements of Section

319 of the federal Clean Water Act. The Unified Watershed Assessment was begun in 1998 to set priorities for restoration of streams where quality had been degraded by a variety of pollution sources *other* than municipal and industrial wastewater treatment plants and discharges. Included among sources are acid mine drainage, sewer system overflows, agricultural runoff, and other nonpoint sources (NPSs) of pollution. This program used PADEP's 305(b) report (PADEP, 2002a) and its 303(d) list (PADEP, 2002b) of impaired streams as a starting point. The PADEP has delineated 104 watersheds that cover the entire state, 30 of which are located in the Ohio

River basin in southwestern Pennsylvania. Each watershed was initially assigned to one of four categories based on the percentage of stream miles assessed, the percentage of these miles judged to be impaired, and the potential for NPS pollution.

Priorities for water quality improvement were assigned to each of the 23 watersheds in Pennsylvania that fall into Category I. Watershed Restoration Action Strategies (WRASs) were then developed for priority watersheds in cooperation with federal, state and local agencies; watershed-based organizations; and the general public. Included among the 30 watersheds in the Ohio River basin for which a WRAS has been prepared are the following: Redbank Creek, Conemaugh River/Blacklick Creek, Stony Creek/Little Conemaugh River, Lower Youghiogheny River, Upper Youghiogheny River/Indian Creek, Upper Monongahela River, Raccoon Creek, and Chartiers Creek.

Each watershed plan includes descriptions of geology and soils, natural and recreational resources, and streams classified by PADEP as being of "exceptional or high quality." Sources of water quality impairment are also discussed. Existing restoration initiatives are listed, and funding needs (to the extent they are known) are estimated. Funding from multiple sources has been provided to address some of the problems covered by these plans. Grants from Pennsylvania Growing Greener, the U.S. Environmental Protection Agency's (EPA's) Section319 and Section 104(b)3, and Pennsylvania's Watershed Restoration Assistance Program have all been received to fund restoration projects. The Pennsylvania Infrastructure Investment Authority also has made loans to local governments to address some of the problems. The PADEP Bureau of Abandoned Mines has also been an active participant in the implementation of many of these watershed plans.

The watershed plans address important issues as identified in Pennsylvania's most recent 305(b) report (PADEP, 2002a) and 303(d) list (PADEP, 2002b) for priority watersheds, but there is no assurance that streams in these watersheds will be restored to a level that fully supports their designated uses. Section 319 requires adoption of best management practices for NPS pollution, but unlike the total maximum daily load (TMDL) process, it does not require a demonstration using predictive models or other evidence that water quality standards will be achieved. Follow-up investigations of projects in WRAS plans will be required to assess progress toward the goal of fully restoring streams in those watersheds.

Contaminated water supplies and improper disposal of sewage from on-site sewage treatment and disposal systems (OSTDSs) not connected to public water or sewer systems were identified in the 2002 WSIP report as being of major concern in the region, but the Unified Watershed Assessment did not include a systematic evaluation of the extent of these problems. As discussed in preceding chapters, better information is needed to make an informed assessment of the locations, magnitude, and priorities to be assigned to these water quality problems.

In contrast to PADEP's WRAS program, which focuses on priority problems within selected watersheds, water supply is being addressed on a basinwide scale that recognizes linkages among watersheds. Pursuant to the WRPA of 2002, PADEP has initiated the process to update the State Water Plan. That act establishes a Statewide Water Resources Committee (SWRC) to set guidelines and policies for the planning process and to conduct a formal review and approval of the product. Regional water resources committees are to be established for each of the state's six major basins. After conducting an open public process and consulting with the SWRC and PADEP, the Ohio Basin Committee is to recommend regional plan components to the SWRC. These areas would be designated, "critical water planning areas," and identified on a multimunicipal watershed basis. Areas in which demand is expected to exceed supplies would be so designated, and more detailed critical area resource plans, or "water budgets," would be established.

The WRPA does not have a similar mandate for water quality. Nevertheless, the planning process it establishes for water supply provides an excellent opportunity for PADEP to exert administrative leadership to better integrate water quality and water supply into a broader framework of planning for water resources at the basin scale. Basin plans should at a minimum indicate the water quality effects on public water supplies and the water quality effects of flood control activities.

Significant legislation enacted by the Pennsylvania General Assembly in 2000 could influence water planning among neighboring local governments. Among other provisions, Pennsylvania Acts 67 and 68 of the 1999-2000 legislative session, Article XI state the following: For the purpose of encouraging municipalities to effectively plan for their future development and to coordinate their planning with neighboring municipalities, counties and other governmental agencies, and promoting health, safety, morals and the general welfare...powers for the establishment and operation of joint municipal planning commissions are hereby granted.

Local governments were given additional powers to regulate growth. Included in those powers were authority to limit development in specially designated "growth areas" and to implement a program of transferable development rights. Municipalities were given authority to enter into intergovernmental cooperative planning and implementation agreements. Municipalities located within the county or counties were also enabled to enter into intergovernmental cooperative agreements to develop, adopt, and implement comprehensive water resource plans for entire counties or any area within counties. Such agreements also enabled participating municipalities to share tax revenues and fees.

The legislation also included incentives for municipalities to enter into such agreements. State agencies were directed (1) to consider multimunicipal plans when reviewing applications for the funding or permitting of infrastructure or facilities, and (2) to consider giving priority to applications for financial or technical assistance for projects consistent with the county or multimunicipal plan. Former Pennsylvania Governor Ridge issued an executive order in January 1999 directing PENNVEST to take land use into consideration when evaluating water project proposals; Acts 67 and 68 of 2000 had similar implications. Among other actions, PENNVEST established as an eligibility requirement that funding of proposed projects be consistent with applicable municipal, multimunicipal, or county comprehensive land use plans and zoning ordinances. How effective this incentive will be in promoting cooperation remains to be seen.

Bibliography

Asaithambi S.: *Economics of Ground Water Management in India*, Abhijeet, Delhi, 2008.

Baecher, G.B.: *Analyzing Exploration Strategies*: *Site Characterization and Exploration*, American Society of Civil Engineers, New York, 1999.

Bagis, Ali Ihsan: *Water as an Element of Cooperation and Development in the Middle East*, Ankara, Ayna Publications, 1994.

Chadha K. L. and Pareek O. P.: *Advances in Horticulture: Fruit Crops,* New Delhi, Malhotra Publishing House, 1993.

Chahal, S.S. :'*Achievements and Prospects in Mycology and Plant Pathology*, International, Delhi, 1997.

Devi, C.R. Sudharmai : *Analytical Procedures in Soil Science and Agricultural Chemistry*, Agrotech, Delhi, 2004.

Dubey, Sarvesh Kumar and Asha Arora: *A Practical Book on Soil, Plant, Water and Fertilizer Analysis*, S.R. Scientific, Delhi, 2010.

Erik Nissen-Peterson : *Rainwater Catchment Systems*, UK: Intermediate Technology Publications, 1999.

Ghosh Biswajit : *Plant Tissue Culture : Basic and Applied*, Universities Press, Delhi, 2005.

Gupta, O.P. : *Water in Relation to Soils and Plants : With Special Reference to Agriculture*, Agrobios, Delhi, 2002.

Helen Bannayan: *Water Resources of Jordan: Present Status and Future Potentials*, Amman, Friedrich-Ebert-Stiftung, 1993.

Herminie Broedel Kitchen: *Soils and Crops: Diagnostic Techniques*, Satish Serial Publishing, Allahabad, 2004.

Jones, R. M.: *Plant Resources of South-East Asia,* Wageningen, Pudoc Scientific Publishers, 1992.

Kanmony, J. Cyril: *Drinking Water Management: Problems and Prospects*, Mittal Pub, Delhi, 2010.

Kapoor, R.L. and M.L. Saini: *Plant Breeding and Crop Improvement*, CBS, Delhi, 1997.

LeRoy, K.: *Sustaining Water Washington*, D.C.: Population Services International, New York, 1993.

Majumdar, S.P. and R.A. Singh: *Analysis of Soil Physical Properties*, Agrobios, Delhi, 2008.

Mathur, S K B B S Kapoor: *Emerging Trends In Soil Management*, Madhu Publications, Delhi, 2007.

Nagamani; A ; I K Kunwar and C Manoharachary: *Handbook of Soil Fungi*, I K International, Delhi, 2006.

Naik, M K and G S Devika Rani: *Advances in Soil Borne Plant Diseases*, New India Pub, Delhi, 2008.

Patel, Sheelwant: *Indian Forests : Soil Water and Bio-Environment Conservation*, Pointer, Delhi, 2005.

Premjit Sharma: *Agricultural Drainage and Water Quality*, Gene Tech Books, Delhi, 2007.

Rao, A.V. Narasimha: *Fundamentals of Soil Mechanics*, Laxmi Publications, Delhi, 2010.

Sathyanarayana B N *: Plant Tissue Culture : Practices and New Experimental Protocols*, I K International, Delhi, 2007.

Shukla, R.S. and P.S. Chandel: *A Textbook of Plant Ecology: Including Ethnobotany and Soil Science*, S. Chand Publisher, Delhi, 2009.

Srivastava, A K and Shyam Singh: *Citrus : Climate and Soil*, International Book Distributing Co, Delhi, 2002.

Swarup, Ram : *Elements of the Nature and Prospectus of Soil*, Manglam Pub, Delhi, 2011.

Terzaghi, K., Peck, R.B., and Mesri, G.: *Soil Mechanics in Engineering Practice*, John Wiley & Sons, Inc., NJ, 1996.

Thapliyal, B K : *Democratisation of Water*, Serials Pub, Delhi, 2008.

Thomas E.: *Fruit Production in St. Kitts and Nevis*, Port of Spain, IICA, 1996.

Tyagi, I.D. : *Plant Breeding and Genetics at a Glance*, South Asian, Delhi, 2005.

Walton, William C.: *Groundwater Resource Evaluation*. McGraw Hill, New York, 1970.

Wilde, S.A.: *Forest Soils and Forest Growth*, Periodicals, Delhi, 1991.

Wood, David Muir: *Soil Behavior and Critical State Soil Mechanics*, Cambridge University Press, Cambridge, 1990.

Index

❑❑❑